SpringerBriefs in Computer Science

Series Editors
Stan Zdonik
Shashi Shekhar
Jonathan Katz
Xindong Wu
Lakhmi C. Jain
David Padua
Xuemin (Sherman) Shen
Borko Furht
V.S. Subrahmanian
Martial Hebert
Katsushi Ikeuchi
Bruno Siciliano
Sushil Jajodia
Newton Lee

More information about this series at http://www.springer.com/series/10028

Dizhi Zhou • Wei Song

Multipath TCP for User Cooperation in Wireless Networks

 Springer

Dizhi Zhou
Alcatel-Lucent Inc.
Ottawa, ON, Canada

Wei Song
Faculty of Computer Science
University of New Brunswick
Fredericton, NB, Canada

ISSN 2191-5768 ISSN 2191-5776 (electronic)
ISBN 978-3-319-11700-3 ISBN 978-3-319-11701-0 (eBook)
DOI 10.1007/978-3-319-11701-0
Springer Cham Heidelberg New York Dordrecht London

Library of Congress Control Number: 2014955146

Printed on acid-free paper

Springer is part of Springer Science+Business Media (www.springer.com)

Preface

Nowadays, the exponential growth of user population and traffic demands poses new challenges to provide quality of service (QoS) with limited wireless resources. User cooperation offers a good opportunity to enable multipath transmission and aggregate the available bandwidths of relays by exploiting the multi-radio capability of mainstream mobile devices. The multipath transport control protocol (MPTCP) by Internet Engineering Task Force (IETF) is a promising multipath solution at the transport layer. In a user cooperation scenario, however, the network conditions become so dynamic and unpredictable that the performance of MPTCP may not be guaranteed. This brief introduces several state-of-the-art extensions to MPTCP that significantly enhance the achievable performance.

Chapter 2 reviews the literature on user cooperation and multipath transmission, followed by an introduction of MPTCP details and extensions to MPTCP. System modelling is discussed in Chap. 3 to characterize MPTCP-based multipath transmission with user cooperation in the Long Term Evolution (LTE) network. Chapter 4 introduces a subset-sum based relay selection (SSRS) module to achieve a stable aggregate throughput with MPTCP in the user cooperation scenario. Two independent and complementary modules, adaptive congestion control (ACC) and differentiated packet forwarding (DPF), are presented in Chaps. 5 and 6, respectively, to improve the goodput based on the stable aggregate throughput provided by SSRS. In Chap. 7, a bandwidth sharing module extends the congestion control algorithm of MPTCP to ensure that the throughput of the local traffic at the relays is not degraded by the forwarding traffic for the destination.

The performance of these extensions to MPTCP is evaluated by ns-3 in a variety of scenarios. The simulation results demonstrate that these modules can achieve a stable aggregate throughput and significantly improve the goodput. Meanwhile, the results show that the extensions can well respect the local traffic of the relays and motivate the relay users to provide the relaying service. Chapter 8 concludes this brief by highlighting possible future research directions.

Ottawa, ON, Canada   Dizhi Zhou
Fredericton, NB, Canada   Wei Song

Contents

Chapter 1
Introduction

This chapter first introduces the motivation of user cooperation and multipath transmission, particulary solutions at the transport layer. Specifically, the multipath transport control protocol (MPTCP) is studied in the user cooperation scenario within the Long Term Evolution (LTE) network. After the discussion of the challenges posed by user cooperation to MPTCP, the state-of-the-art multipath solutions that this brief presents are briefly outlined.

1.1 Motivations

1.1.1 User Cooperation

Nowadays, the fast development of wireless communication technologies provides a good opportunity for mobile devices to run a variety of bandwidth-intensive applications, such as video streaming and video conferencing[8]. By adopting cutting-edge technologies, e.g., multiple-input and multiple-output (MIMO) and orthogonal frequency-division multiplexing (OFDM), the bandwidth and transmit rate of the cellular network are increased dramatically. Long Term Evolution (LTE) can already support a maximum downlink peak rate of 300 Mbps and an uplink peak rate of 75 Mbps to mobile users [1]. Meanwhile, Wi-Fi hotspots are largely deployed in both outdoor and indoor places, such as the airport, university campus, coffee house and business building. These hotspots usually offer a higher transmission rate than the cellular network. For instance, the latest IEEE 802.11n can support a data rate of 600 Mbps with the use of four spatial streams at a channel bandwidth of 40 MHz [5].

All these high capacity wireless networks enable applications that require a high bandwidth in mobile devices. A report by Cisco in 2010 shows that almost 70 % of mobile traffic will carry video data in 2015 [2]. Providing a stable quality

© The Author(s) 2014   1
D. Zhou, W. Song, *Multipath TCP for User Cooperation in Wireless Networks*,
SpringerBriefs in Computer Science, DOI 10.1007/978-3-319-11701-0_1

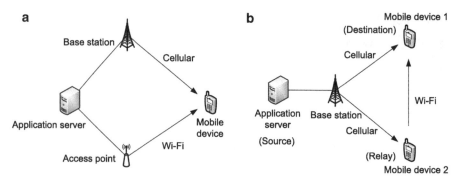

Fig. 1.1 Cooperations in wireless networks. (**a**) Network cooperation. (**b**) User cooperation

of service (QoS) required by mobile video users is still, however, a challenging issue in wireless networks. There are several reasons behind this problem. First, wireless networks cannot guarantee a full coverage with the same signal strength to all mobile users. In some indoor environments, such as inside an elevator or a basement, the signal strength of the cellular network is much lower than that in outdoor places. Second, more importantly, in both cellular and Wi-Fi networks, the bandwidth available to an individual user is often much lower than the peak rate advertised by wireless network providers, due to the fact that many users compete for wireless resources at the same time. Actually, in a real deployment, the rate that a user can achieve depends on many different factors, such as the distance to the base station (BS), the number of users simultaneously connected to the BS, and the bandwidth of the backhaul connection from the associated BS to the wired Internet. All these uncertain factors can impact the stable QoS required by mobile users.

In 2010, PC Magazine conducted a national test in the U.S. and found that most wireless networks could not achieve their promised service rates. For example, the high speed packet access plus (HSPA+) network of T-Mobile in Philadelphia only provided a low downlink rate in the downtown center, which was one-fifth of that in rural areas [6].

Network cooperation is one approach to address the above problem, which is shown in Fig. 1.1a. Nowadays, mainstream mobile devices are usually equipped with multiple radio interfaces. Such multi-radio mobile devices often have at least one built-in wireless wide area network (WWAN) interface, e.g., LTE, as well as one or more short-range wireless network interfaces, e.g., Wi-Fi and Bluetooth. Therefore, a multi-radio mobile device can connect to more than one base station in different wireless networks using multiple interfaces. By this means, the bandwidths of multiple wireless networks can be aggregated so as to enhance the QoS for the mobile user.

Such a network cooperation approach, however, is subject to a high power consumption for one mobile device due to running multiple interfaces simultaneously. Meanwhile, not all wireless networks have ubiquitous coverage. For example, Wi-Fi networks are often deployed in disjoint hotspots. When there is no Wi-Fi coverage (e.g., within subways), only the WWAN interface can be used. In addition, even though there can be multiple WWAN interfaces installed in one device, most mobile devices only have one subscriber identity module (SIM) card so that it can only connect to one wireless network at one time. All these problems largely limit the application of the network cooperation solution.

User cooperation by pooling mobile devices in the same vicinity together as a user cooperation group [7] is an alternative way to provide a stable QoS to mobile users. A user cooperation scenario is shown in Fig. 1.1b, where mobile device 2 can receive packets on behalf of mobile device 1 via its own cellular interface and then forward the packets toward mobile device 1 via short-distance communications (e.g., Wi-Fi). In this way, mobile device 1 can aggregate the bandwidth of nearby mobile device 2 so as to provide a larger bandwidth to its applications. Actually, mobile device 2 serves as a relay for mobile device 1. This brief refers to the mobile device that offers the relaying service to others as the *relay*, e.g., mobile device 2 in Fig. 1.1b. Meanwhile, the brief refers to the mobile device that receives packets and utilizes the relaying service provided by others as the *destination*, e.g., mobile device 1 in Fig. 1.1b. Also, the brief refers to the node that sends packets to the destination as the *source*, e.g., the application server in Fig. 1.1b. Compared to the network cooperation solution, the user cooperation approach has many unique benefits. First, the power consumption of the destination is balanced by nearby relays that receive packets for it. Second, even if the destination can only connect to the cellular network, e.g., when a Wi-Fi hotspot is not available nearby, the destination can still enhance its bandwidth by connecting to relays using its Wi-Fi interface. As seen, compared to the network cooperation solution, the user cooperation approach can be applied in many more scenarios.

One of the key issues in user cooperation is the motivation for nearby mobile devices to forward packets for others. In recent years, an increasing number of users own multiple mobile devices, such as a smartphone, laptop, tablet PC and E-reader. According to the survey by GSMA Intelligence, a user in the U.S. owns 1.57 mobile devices on average in 2013 [3]. Most of mobile devices now have both Wi-Fi and cellular interfaces installed. Therefore, the destination and relay devices in one vicinity may belong to the same user or several users who set up a user cooperation group. As a result, the involved mobile users are motivated to help each other receive and forward packets, while benefiting from the relaying service of others. For example, when a user is watching an online HD video by iPad, he or she can utilize the bandwidth provided by his or her friends' iPhone.

Given the advantages and flexibilities of user cooperation, many protocols are designed and proposed for the user cooperation scenario so as to maximize various benefits. This brief will present several state-of-the-art solutions that extend the transport layer protocols to accommodate specific characteristics of user cooperation and maximize the user achievable performance at the destination.

1.1.2 Multipath Transmission

As discussed in Sect. 1.1.1, bandwidth-intensive applications, e.g., video streaming, will become the main stream of traffic in wireless networks. Usually, these applications require a large bandwidth and are easily affected by a network failure (e.g., due to the loss or severe degradation of the radio signal). Therefore, the future wireless network should not only offer a high network capacity, but also provide always best connectivity to mobile users [15].

Traditionally, the standard IP-based protocols deliver packets between two end nodes along a single transmission path. Although the single path may be adjusted and modified by some intermediate routers, the communicating nodes can only utilize one path at one time. Such protocols that rely on a single path cannot adapt well to the highly dynamic environment of wireless networks. In a cellular network, channel fading, path loss and signal interference can degrade the wireless link quality, which can further result in varying throughput to mobile users. Even worse, it is not easy to predict these factors accurately. As a result, the performance of single-path protocols can be seriously affected without appropriate configuration based on accurate estimates. Even though some single-path protocols can migrate traffic to a better path so as to avoid a low quality link, these protocols require additional handover time and signal overhead.

Another approach that can address the problems of single-path protocols is to simultaneously use multiple paths between two communication nodes. Such multipath transmission offers many benefits. First, the achievable throughput between two nodes can be increased by aggregating the bandwidth of multiple paths. Second, when one path is disconnected, the connection can still be maintained by other available paths. Third, the multipath transmission can be used to distribute the traffic load over different paths. When the throughput on one path becomes lower, the source can migrate its traffic to other paths having a larger bandwidth to ensure a stable aggregate throughput.

These potential benefits of multipath transmission lead to many multipath protocols being proposed for different layers to satisfy different objectives. First, the link layer multipath solution mainly focuses on the bandwidth aggregation within the local network. Such solutions cannot select paths in a broader perspective (e.g., engaging different types of links) because their application scenario is limited. For example, for the user cooperation scenario in Fig. 1.1b, the relay receives packets from the base station via the cellular link and forwards the packets to the destination via the Wi-Fi link. Obviously, the link layer multipath solution cannot take full advantage of such a heterogeneous environment.

Second, the network layer multipath solution mainly aims at smart and failure-tolerant routing. By maintaining multiple connections during the handover in the wireless network, mobile users can be always-connected even when they move from the coverage of one base station to another. The network layer multipath protocols cannot, however, meet some key requirements of future wireless networks. For example, they cannot balance the traffic load automatically since the network layer is not able to detect the congestion on a path. As a result, these network layer multipath protocols must rely on additional load balancing solutions, which will introduce extra overhead. Moreover, because the packets that arrive at the receiver can be out-of-order due to the various end-to-end delays of different paths, the performance of the upper layer protocols, e.g., the transport control protocol (TCP), can be seriously jeopardized.

Third, the transport layer multipath solutions attract more and more attention in recent years due to some unique advantages. In addition to the benefits of bandwidth aggregation and always connected state offered by the network layer solutions, the most attractive feature of the transport layer multipath protocols is the awareness of congestion on the path. A well-designed transport layer multipath congestion control algorithm can not only aggregate the bandwidths of multiple paths, but also avoid harming other single-path flows [9]. In addition, the out-of-order problem that presents at the network layer can be easily solved by using the sequence number of the connection-oriented transport layer protocols (e.g., TCP).

Some multipath transmission protocols have been proposed for the application layer, e.g., for peer-to-peer (P2P) applications. Such protocols are mainly focused on specific applications and cannot be used for general purposes.

Jointly considering the strengths and limitations of implementing multipath transmission at different layers, this brief focuses on the transport layer solutions, aiming particularly at the user cooperation scenario. Such an application scenario poses some unique challenges to enable multipath transmission at the transport layer.

1.2 Challenges

As discussed in Sect. 1.1.2, multipath transmission at the transport layer has many unique benefits compared to other layers, such as the awareness of path congestion and fairness to single-path flows. As a possible solution, the multipath transport control protocol (MPTCP) [4] was made an Internet draft by Internet Engineering Task Force (IETF) in 2011. MPTCP runs in multi-radio mobile devices to deliver packets simultaneously over multiple paths via different radio interfaces.

As a multipath transmission protocol at the transport layer, MPTCP is designed to maximize the aggregate throughput, balance the traffic load among paths and ensure the fairness to single-path TCP flows. In the user cooperation scenario of wireless networks, network conditions are more dynamic and unpredictable than in the wired environment. In order to provide stable QoS to the upper layers, MPTCP needs to address several critical challenges.

First, the available bandwidth provided by a relay is highly dynamic due to the fading effect of the wireless channel and the varying local traffic load, which may in turn introduce serious side effects to bandwidth-intensive applications. The term, *available bandwidth*, is used to refer to the bandwidth that a relay provides to the destination. An essential problem arises, given distinct and varying available bandwidths of multiple relays, how does MPTCP guarantee a stable aggregate throughput to the application layer of the destination?

Second, a stable aggregate throughput provided by MPTCP is not sufficient to satisfy the stringent QoS requirements of applications because it is the goodput that reflects the real application-level requirement. Here, goodput is defined as the amount of useful data available to the receiver application per unit time. In other words, goodput represents how many in-order packets received at the transport layer can be delivered to the application layer per unit time. The available throughput via each relay may be varying to such a large scale that the end-to-end delay of each path can be considerably different. Disparate end-to-end delays can cause out-of-order packets received at the destination and thus jeopardize the goodput at the destination. As seen, even when the same aggregate throughput is provided by MPTCP, the user perceived QoS can vary significantly due to the different achieved goodput.

Third, MPTCP needs to ensure that the multipath flow does not harm the local traffic of relays. Even when MPTCP engages the relays for forwarding packets to the destination, the relays should be guaranteed the same throughput for the local traffic as that when they do not help forward traffic. Otherwise, mobile users would not be motivated to provide any relaying service. Unfortunately, it is found in this work that MPTCP may not meet this requirement when the sending rates of some local flows at the relays are greater than an expected fair share.

1.3 Scope

In the literature, there have been many studies on user cooperation and multipath transmission. Particularly, considering the challenges detailed in Sect. 1.2, this brief presents several state-of-the-art solutions that extend MPTCP to achieve a high multipath transmission efficiency with user cooperation in the LTE network.

First, the *subset-sum based relay selection* (SSRS) module proposed in [12] is introduced as a representative solution at the destination that can guarantee a stable aggregate throughput that satisfies the application-layer *target bit rate* (TBR) requirement of the destination. The TBR can be the minimum or desired bandwidth requirement of specific applications. The background traffic at the relay can be based on user datagram protocol (UDP), TCP or more generic additive-increase and multiplicative-decrease (AIMD) control.

Second, this brief introduces a proactive module proposed in [13], referred to as *adaptive congestion control* (ACC), for the source to achieve similar end-to-end delays over multiple paths so that the number of out-of-order packets can be reduced. ACC can enhance the achievable goodput of the destination, so that the performance gain from the transport layer, which is the stable aggregate throughput, is seamlessly transferred to the application layer. Nonetheless, ACC cannot eliminate the end-to-end delay difference of paths, since many factors of delay, such as the queue length at routers and retransmission over wireless links, are inevitable. Therefore, a reactive module proposed in [10], referred to as *differentiated packet forwarding* (DPF), is presented to complement ACC. DPF works at the destination and relays. It temporarily buffers out-of-order packets at the relays so as to improve the goodput at the destination. The two modules work independently and also are mutually complementary with each other.

Third, bandwidth sharing for multipath transmission in a user cooperation scenario is investigated in [11,14]. This brief presents two extensions to the MPTCP congestion control (MCC) algorithm to ensure that the multipath flow for the destination does not degrade the throughput of the local traffic of relays. This is to keep the relays motivated to forward packets to the destination. The first extension, referred to as *MCC-Coop*, aims to ensure that the MPTCP flow of the destination runs fairly with the local single-path TCP flows of relays. To further protect local AIMD flows, another more generic extension, referred to as *GMCC-Coop*, is also discussed.

The above extension modules to MPTCP can work together in a complementarily fashion. The SSRS and bandwidth sharing schemes can be viewed as the foundation for MPTCP to provide a stable aggregate throughput and respect the local traffic of relays. Additionally, ACC and DPF can further take full advantage of relays to maximize the application-level goodput, which is one of the key indicators for the user-perceived QoS. Before more details are introduced about these solutions, Chap. 2 will first review related research issues in the literature, followed by system modelling in Chap. 3.

References

1. Astely, D., Dahlman, E., Furuskar, A., Jading, Y., Lindstrom, M., Parkval, S.: LTE: The evolution of mobile broadband. IEEE Commun. Mag. **47**(4), 44–51 (2009)
2. Cisco: Cisco visual networking index: Forecast and methodology: 2012–2017. http://www.cisco.com/en/US/solutions/collateral/ns341/ns525/ns537/ns705/ns827/white_paper_c11-481360_ns827_Networking_Solutions_White_Paper.html (2013)
3. Fitchard, K.: The average US subscriber owns 1.57 mobile devices. http://gigaom.com/2012/10/22/the-average-us-subscriber-owns-1-57-mobile-devices/ (2013)
4. Ford, A., Raiciu, C., Handley, M., Barre, S., Iyengar, J.: Architectural guidelines for multipath TCP development. IETF RFC 6182 (2011)
5. Perahia, E.: IEEE 802.11n development: History, process, and technology. IEEE Commun. Mag. **46**(7), 48–55 (2008)
6. Phonescoop: 4G networks tested: WiMAX vs. HSPA+. http://www.phonescoop.com/articles/article.php?a=376&p=2707 (2013)
7. Sharma, P., Lee, S., Brassil, J., Shin, K.: Aggregating bandwidth for multihomed mobile collaborative communities. IEEE Trans. Mobile Comput. **6**(3), 1536–1233 (2007)
8. Song, W., Zhuang, W.: Performance analysis of probabilistic multipath transmission of video streaming traffic over multi-radio wireless devices. IEEE Trans. Wireless Commun. **11**(4), 1554–1564 (2012)
9. Wischik, D., Raiciu, C., Greenhalgh, A., Handley, M.: Design, implementation and evaluation of congestion control for multipath TCP. In: Proc. USENIX NSDI (2011)

10. Zhou, D., Ju, P., Song, W.: Performance enhancement of multipath TCP with cooperative relays in a collaborative community. In: Proc. IEEE PIMRC (2012)
11. Zhou, D., Song, W., Cheng, Y.: A study of fair bandwidth sharing with AIMD-based multipath congestion control. IEEE Wireless Communications Letters **2**(3), 299–302 (2013)
12. Zhou, D., Song, W., Ju, P.: Subset-sum based relay selection for multipath TCP in cooperative LTE networks. In: Proc. IEEE GLOBECOM (2013)
13. Zhou, D., Song, W., Shi, M.: Goodput improvement for multipath TCP by congestion window adaptation in multi-radio devices. In: Proc. IEEE CCNC (2013)
14. Zhou, D., Song, W., Wang, P., Zhuang, W.: Multipath TCP for user cooperation in LTE networks. IEEE Network (2014). To appear
15. Zhuang, W., Mohammadizadeh, N., Shen, X.: Multipath transmission for wireless Internet access - From an end-to-end transport layer perspective. Journal of Internet Technology **13**(1) (2012)

Chapter 2
User Cooperation and Multipath Transmission

This chapter first reviews the literature on user cooperation at different layers of the network protocol stack. Then various issues of multipath transmission at the transport layer are discussed. After that, the protocol details of the IETF solution, MPTCP, are introduced. The related studies on MPTCP are surveyed at the end of this chapter.

2.1 User Cooperation

In the past decade, there have been extensive studies on user cooperation at different layers. This section reviews the important research results on user cooperation.

2.1.1 Physical Layer

User cooperation at the physical layer is often termed as *cooperative communications*, which allows single antenna mobile devices to reap the benefits of MIMO systems. The basic idea is that the single antenna mobile devices can share their antennas with other mobile devices by creating a virtual MIMO system [47]. In such a cooperative communication scenario, the mobile devices can only use a single antenna to share the transmission capacity with others over the same type of wireless links.

The theoretical foundation of cooperative communications can be traced back to the work of Cover and El Gamal in [12]. The capacity of the cooperative system, including a source, a destination and a relay mobile device, is analyzed by assuming that all the above nodes run in the same frequency band. Based on this analysis, many cooperation protocols are proposed so as to achieve a large wireless

© The Author(s) 2014
D. Zhou, W. Song, *Multipath TCP for User Cooperation in Wireless Networks*,
SpringerBriefs in Computer Science, DOI 10.1007/978-3-319-11701-0_2

network capacity by incorporating relays in the transmission between the source and destination. Specifically, while the source transmits the signal to the destination, the relay can forward the received signal from the source by different schemes. In the rest of this section, some representative solutions are reviewed.

Amplify-and-Forward

This method is proposed by Laneman et al. in [38]. As the name implies, the relay in this method receives a noisy version of the signal from the source. Then, the relay simply amplifies the signal and retransmits it to the destination. The simplicity and cost-effectiveness are two main advantages of the amplify-and-forward method. Many studies are done to analyze its performance in various wireless environments [30]. In [4], Anghel and Kaveh consider a wireless system with K relays and derive the average symbol error rate at the destination in the Rayleigh fading channel. The K-relay scenario is very useful in practice since a single relay may not make enough contribution to improve the capacity. The multi-hop amplify-and-forward relay scenario is investigated by Ribeiro et al. in [58] and a relatively accurate approximation is derived for the symbol error probability. These amplify-and-forward solutions assume that the base station knows the inter-user channel coefficients, so additional signalling is required to exchange such information in the implementation [47].

Decode-and-Forward

In this method, the relay decodes the source message and retransmits the re-encoded message to the destination [62, 63]. Compared to the amplify-and-forward method, decode-and-forward is more complex. Some solutions are proposed to balance the performance and complexity in the real system design. In [25], Hsu et al. propose a decode-and-forward solution for an OFDM network so as to maximize the system weighted sum rate. Specifically, working in the half-duplex mode, the relay decodes the message on a particular subcarrier at one time slot, and then re-encodes and forwards the fresh message on a different subcarrier at the next time slot. Therefore, each message is transmitted on two subcarriers in two hops. There are also some studies on extending decode-and-forward into a multi-relay scenario. The work in [15] addresses the relay selection for decode-and-forward with multiple relays based on the symbol error rate.

Compress-and-Forward

In this method, the relay quantizes and compresses the received signal from the source, and retransmits the encoded signal to the destination [35]. Similar to amplify-and-forward and decode-and-forward, compress-and-forward is also based on the theoretical work of [38]. So far, compress-and-forward has been considered for implementation in many wireless systems. In [69], the authors propose a compress-and-forward scheme for MIMO-OFDM transmission. Specifically, the capacity of the compress-and-forward scheme is analyzed for the time division duplex (TDD) scenario. Simoen et al. further derive the achievable rates over Gaussian vector channels with compress-and-forward relaying [68].

Coded Cooperation

In coded cooperation, channel coding is introduced into cooperative communications [27, 28]. In a different approach from the previous cooperation methods, the relay of the coded cooperation does not repeat the signal received from the source. Instead, the source divides the data into two blocks which are augmented by the cyclic redundancy check (CRC) code. The source transmits different blocks to the relay and the destination via two independent fading paths. The main idea of coded cooperation is that the source tries to transmit incremental redundant blocks to its relays. Different channel coding methods can be used in coded cooperation [47]. Bao and Li propose a generalized adaptive network coded cooperation scheme, which utilizes varied channel coding based on different network environments [6]. In [45], a coded cooperation scheme based on the Turbo code is proposed for the environment of high signal-to-interference plus noise ratio (SINR).

The above four methods are typical solutions to enable user cooperation at the physical layer. Specifically, amplify-and-forward, decode-and-forward and compress-and-forward are based on the theoretical work of [38]. Coded cooperation is a combination of channel coding and cooperative communications. These methods are used in various types of wireless networks (e.g., OFDM) and scenarios (e.g., multiple relays and multiple hops). Physical layer cooperation, however, focuses on the same type of wireless links. Hence, these techniques cannot be directly applied to aggregate the network capacity in a heterogenous environment.

2.1.2 MAC Layer

The traditional MAC layer mainly aims to coordinate multiple nodes for sharing the wireless medium [32]. There are two basic categories of MAC protocols: channel allocation and contention-based random access. In channel allocation based MAC, the base station divides wireless resources based on different dimensions, such as time division multiple access (TDMA), frequency division multiple access (FDMA), and code division multiple access (CDMA). On the other hand, contention-based random access shares the resource based on the competition among mobile devices. Example solutions in this category include ALOHA and carrier sensing multiple access with collision avoidance (CSMA/CA) in IEEE 802.11 [57].

In the user cooperation scenario, the role of relays must be further considered in the wireless medium access schemes. Furthermore, a cooperative MAC protocol needs to address two key issues [65, 81]:

- When to use cooperation?
- Whom to cooperate with?

Many cooperative MAC protocols are proposed to address the two issues for the user cooperation scenario. In [78], the authors propose two channel allocation mechanisms with different complexities for a cooperative cognitive radio network to maximize the achievable end-to-end throughput. In [76], Yang et al. propose a cooperative TDMA scheme to enable user cooperation over the Rayleigh fading channel so as to enhance the correct packet reception probability and the system capacity. The base station manages the cooperation between the source and the relays. These two schemes are typical solutions based on channel allocation. Due to the complexity concern with channel management, the contention-based cooperative MAC protocols attract more attention. The survey in [32] reviews the representative contention-based solutions and classifies them in the following categories.

Category I

In this category, the source decides when to use cooperation, and how to select the relay candidates and the optimal relays [29, 42, 79]. In the MAC protocol proposed in [29], the source determines when it needs to cooperate with the relays and selects the best relay based on the available partial channel state information (CSI). Liu et al. [42] follow a similar idea but further design the CoopMAC protocol, which specifies the data structure, message sequences and formats. Such design details are essential for the real implementation of cooperative MAC protocols.

Category II

In this category, it is still the source that decides when to use cooperation and how to select the optimal relay. The relay candidate list is, however, built based on the competition among the relays [50, 80, 82]. Specifically, the authors of [80] propose an algorithm, referred to as relay-enabled distributed coordination function (rDCF), to help the source, relay and destination to achieve an agreement on the relay selection in a distributed approach. In rDCF, a relay periodically broadcasts messages to show its willingness to cooperate, whenever it detects that it can improve the transmission rate between the source and destination. Once a relay hears that more than M relays are willing to help with the same source and destination pair, it stops sending broadcast messages. As such, the source can obtain the relay candidate list through the relay contentions. Zou et al. propose an enhanced scheme for rDCF, in which the relay can only start to broadcast messages when other relays stop broadcasting and are in a sleeping mode [82].

Category III

In this category, the source only decides when to use cooperation. In contrast, the relay is selected by a source in a distributed manner rather than in a centralized manner [8, 64]. Bletsas et al. propose an opportunistic relay selection scheme [8]. All relays estimate the instantaneous channel conditions and maintain a timer. Usually, the timer of the relay with the best channel quality will expire first. Hence, the source can select the best relay which is the first node that broadcasts a message to claim its willingness for cooperation. In [64], a better relay is indicated by less channel access time. Then the best relay is the first node that responds to the source. Therefore, there is no need for the relay to broadcast messages to other competitors, which alleviates the network load from broadcast traffic. One common requirement for the timer-based protocols is that appropriate synchronization is needed among the relays in the user cooperation scenario [32].

Similar to the user cooperation at the physical layer, the mobile nodes involved in the cooperation at the MAC layer also utilize the same type of wireless links.

2.1.3 Network Layer

The network layer is responsible for packet routing through intermediate routers. In the user cooperation scenario, there can be multiple transmission paths from the source to the destination through different relays. Therefore, a multipath routing scheme can be used at the source and destination so as to achieve a large aggregate bandwidth and a high degree of tolerance to transmission failures.

One key problem in multipath routing is to find multiple available paths between the source and destination. In [81], a multipath route is classified into the following three categories:

- Node disjoint route: there are no common nodes or links among multiple paths between the source and destination;
- Link disjoint route: there are no common links among multiple paths between the source and destination. However, there are common nodes among the paths; and
- Non-disjoint route: there are common nodes or links among multiple paths between the source and destination.

According to this classification, the disjoint route can provide the most aggregate resources and the highest degree of fault tolerance. Finding the disjoint route is difficult because, in many cases, the disjoint route is also the optimal route [46]. The problem of finding an optimal route with some additional constraints is usually NP-hard. The example in [81] for multipath routing aims to find the paths whose aggregate bandwidth meets a required value and whose largest delay is smaller than a threshold. As this problem is NP-hard, many multipath routing protocols use a heuristic approach to find the disjoint routes [21, 75].

Relay selection is another key issue of multipath routing in the user cooperation scenario. The authors of [67] propose a multipath routing solution to select the relays in a multihop wireless network. In [19], the source selects an optimal relay set so as to minimize the energy consumption with user cooperation in wireless sensor networks.

Unfortunately, the network layer multipath protocols cannot meet some key requirements of future wireless networks. For example, they cannot balance the traffic load automatically since the network layer is not able to detect the congestion on a path. Additionally, because the packets that arrive at the receiver can be out-of-order due to the various end-to-end delays of different paths, the performance of the upper layer protocols, e.g., the transport control protocol (TCP), can be seriously affected.

2.1.4   Transport Layer

The performance of the transport layer protocol can be affected by engaging user cooperation at the lower layers, e.g., the out-of-order problem of the multipath routing at the network layer. There are many solutions proposed to improve the performance of the transport layer protocols in a user cooperation scenario.

Most of these solutions follow the principle of cross-layer design, which jointly considers the parameters at the lower layers and the transport layer. Kwasinski studies the TCP performance with user cooperation based on decode-and-forward or amplify-and-forward at the physical layer in [37]. Packet retransmission at the MAC layer is also used to further improve the TCP throughput.

Although many solutions assume that the relays are given and known, the relays selected by the lower layers may not translate the performance gain to the transport layer. Hence, it is worth studying the relay selection at the transport layer for the user cooperation scenario. In [72, 73], the authors propose a cross-layer relay selection solution to optimize the TCP throughput in a cooperative relaying network. The best relay is determined by a function of SINR at the physical layer, and the frame size and retransmission time at the MAC layer. Hu and Li design an energy efficient relay selection scheme for TCP by jointly considering the lower layer parameters, e.g., the frame size and the maximum retransmission time [26]. The relay selection algorithm proposed by Chen et al. in [11] jointly considers power allocation, adaptive modulation and coding, and the frame size at the MAC layer to maximize the TCP throughput in a cognitive relay network.

In [34, 66], the authors propose a multipath transmission protocol by using multiple TCP flows in the user cooperation scenario. The source selects the optimal relay set based on the link state of each candidate relay, and forwards encapsulated packets to the relays via a generic routing encapsulation (GRE) tunnel. Then the relay decapsulates and forwards the packets to the destination via the Wi-Fi link. After that, the destination returns the ACK packets to the source via its own WWAN link. The source can avoid unnecessary slowing down of the sending rate by using the out-of-order packet information contained in these ACK packets.

2.1.5 Application Layer

User cooperation at the application layer usually aims to improve the performance for a specific type of application. The existing solutions can be classified into the following categories.

Video Streaming

CStream [61] is a user cooperation solution at the application layer for video streaming. In CStream, the destination first broadcasts the relay request to nearby nodes to setup a user cooperation group. Then the destination sends the list of selected relays to the source. The source creates a buffer queue and a thread for each relay. Each thread fetches a packet from the source when it finishes sending a packet to the relay. Once the relay receives a packet from the source, it forwards the packet to the destination immediately. If the relay fails to forward the packet, the destination can detect the failure by monitoring the I-CAN-HELP message sent from the relay periodically. The destination also periodically sends a relay update message to the source so that the source can schedule the packet to new relays.

Different from the scenario in CStream, there are many cooperation protocols for video streaming, which assume that the destinations request the same video resource from the application server. For instance, a group of mobile users in the same vicinity watch a broadcast TV channel at the same time. The authors of [56] evaluate the power consumption for both the non-cooperation and cooperation scenarios. Instead of sending the whole video file independently to each destination, the source divides the video file into several partitions and sends each partition to a destination via the Wi-Fi link. If there are N destinations requesting the same video, the source will divide the file into N pieces and send the packets of one piece to one destination in a round-robin fashion. Then these destinations exchange the packets from the source between each other through its Bluetooth link. As such, these destinations act as relays for each other. The test results show that the power consumption of the destinations can be notably reduced without compromising the video quality by using user cooperation.

The protocol proposed in [2] follows a similar approach, but focuses on a different video compression algorithm—H.264/SVC, which is a video compression standard that has a strong video compression capability [60]. H.264/SVC divides the video data into a base layer and multiple enhancement layers. The base layer provides the basic video quality, while the enhancement layer can be used to enhance the achieved quality. In [2], the source broadcasts the base layer data to the destinations to guarantee the basic video quality. Meanwhile, the source sends different enhancement layer data to different destinations via the WWAN link. Then these destinations transfer the enhancement layer data from the source to each other through a Wi-Fi link. The proposed approach in [36] uses a base station with multiple antennas to send out H.264/SVC video data encoded by space-time code. The destination that does not cooperate with others can only decode the base layer data. For the cooperative destinations, the enhancement layers can be exchanged among each other and the visual quality can be improved.

Different from the above solutions, only part of selected destinations act as relays in [44]. For instance, within a user cooperation group of N destinations, one destination is selected as the on-duty relay to receive and forward broadcast video for others. Several backup relays are also selected to monitor the status of the on-duty relay. These backup relays can easily take over the role of the on-duty relay, once the on-duty relay leaves the user cooperation group. If a new destination joins the user cooperation group, the on-duty relay will send the latest video data to it. This approach allows the new destination to start playing out fast, since it does not need to wait for the next broadcast burst.

The authors of [43] propose a network-coding-based data repair framework for the user cooperation scenario so as to improve the broadcast video quality. The network coding scheme is used when the destinations exchange the video data from the source with each other so that the packet recovery ability can be enhanced.

File Downloading

COMBINE is a user cooperation solution at the application layer for file download-ing via the hypertext transfer protocol (HTTP) [3]. In the user cooperation scenario, the destination first assigns an amount of chunks to each relay based on the available bandwidth of the relay. Then, the relays download different parts of the file via the WWAN link, and forward the packets to the destination via the Wi-Fi link after downloading all chunks assigned to it. If the relay has its own traffic or fails to transmit the packets (e.g., if the relay is powered off or moves out of the transmission range), the destination will select another relay to download the packets. In order to maintain the HTTP session, COMBINE implements a Web proxy in the network side so that the destination and relays can be shown to have a single IP address to the source. The proxy greatly enhances the compatibility of the cooperative protocol and reduces the response time to recover from network failures.

Pricing

Pricing is one key issue of user cooperation at the application layer. A fair pricing scheme can motivate mobile users to offer the relaying service to others. In COMBINE [3], the authors design an accounting scheme that each relay can broadcast its relaying service's price to others. Once the destination selects one or multiple relays based on the price, it sends the response to these relays to form a user cooperation group and then sends the download mission to each relay. The relaying price is calculated by the monetary and energy costs of the relaying service.

The authors of [10] propose an open market architecture, referred to as mobile bazaar (MoB), in which the destination and relays can flexibly trade various services with each other. For example, MoB considers the packet forwarding as a service which can be advertised by the relays and discovered by the destinations. MoB mainly focuses on the service billing, reputation and security issues in the user cooperation scenario.

2.2   Multipath Transmission at Transport Layer

Multipath transmission at the transport layer can address some issues caused by the multipath routing protocol at the network layer. As shown in [81], a multipath transport layer protocol needs to address the following issues:

- Multi-homing capability;
- Simultaneous transmissions;
- Path assignment; and
- Packet reordering.

In order to support multipath transmission, the source and the destination must maintain multiple IP addresses. Traditional transport layer protocols, such as TCP and UDP, can only support a single IP address. Although the source and the destination can establish multiple TCP/UDP connections, it is not fair to the single-path flow because a multipath flow occupies much more bandwidth. In addition, the goodput in this case may be much lower than the aggregate throughput, since each connection works independently and the unmatched rates of different paths will cause many out-of-order packets. Therefore, it is necessary to design an effective multipath protocol at the transport layer to support a multi-homing capability.

Stream control transmission protocol (SCTP) [70] is an IETF standard which provides the multi-homing capability to a node with multiple IP addresses. It defines a transport layer connection as an association. Multiple IP addresses are bound to an association so that the source can utilize multiple IP addresses to communicate with the destination. Another standard solution for the multipath transmission at the transport layer is multipath TCP (MPTCP) [17], which runs in a multi-homed node to simultaneously deliver TCP packets over multiple paths.

Both SCTP and MPTCP support path assignment and packet reordering. For the path assignment, the source needs to determine which path to send on for each packet. Various factors can be considered in the decision criteria, such as bandwidth, round-trip time (RTT), packet loss rate and so on [24].

On the other hand, the packet reordering problem can be caused by different end-to-end delays of multiple paths. Out-of-order packets can trigger the destination to send back selective acknowledgement (ACK) messages which will be interpreted as missed packets by the source. Then the source will unnecessarily retransmit the packets and may slow down the sending rate. Both SCTP and MPTCP have mechanisms to avoid such action. However, different from MPTCP, SCTP aims to use simultaneous transmission for failure tolerance. An additional path can only be activated when the original path is disconnected or when some packets need to be retransmitted. Several extensions have been proposed to SCTP to extend its functionality [1, 31, 41]. The authors of [31] propose a concurrent multipath transfersolution based on SCTP so as to enable simultaneous multipath transmission.

SCTP is not widely deployed due to the lack of support for the middle box, e.g., the network address translation (NAT) box. In contrast, MPTCP attracts more attention in recent years because it can be run on the existing network protocol stack and easily traverse the middle boxes on the path. For example, the latest mobile operating system iOS7© has already adopted MPTCP for the traffic generated by Siri© [9]. Actually, MPTCP extends the regular single-path TCP to add multipath capability. There are also other TCP-based multipath transport protocols. They mainly need to deal with two key issues, which are the extensions of the protocol structure and the congestion control algorithm.

2.2.1 Protocol Structure

In order to support multipath transmission at the transport layer, many multipath protocol structures are proposed to aggregate bandwidth over multiple end-to-end paths [23, 77]. Hsieh and Sivakumar propose the pTCP protocol to bind multiple paths regardless of the characteristics of an individual path [23]. pTCP modifies and adds several key modules into the existing protocol stack, which include the multihoming capability, service differentiation using a purely end-to-end mechanism and so on. As these protocols modify the structure of the existing protocol stack, they are not widely deployed in practice.

On the other hand, some protocols utilize multiple single-path TCP flows and add some additional mechanisms to efficiently aggregate the throughput of multiple paths. The authors of [71] propose an algorithm of using multiple TCP connections by adjusting the receiver window size of each single-path TCP flow to achieve the desired throughput for multimedia streaming. Although these protocols keep the standard protocol stack, using multiple single-path TCP flows simultaneously cannot guarantee TCP-friendliness, which is another key requirement for the transport layer support for multipath transmission.

2.2.2 Congestion Control

A well-designed congestion control algorithm for the multipath transport protocol should not only efficiently aggregate bandwidth and balance traffic among paths, but also compete for bandwidth fairly with single-path TCP flows, which is the TCP-friendly requirement. Even though the source and the destination can establish multiple TCP connections, it is not fair to the single-path TCP flow, which occupies relatively less bandwidth. Honda et al. propose the equally-weighted TCP (EWTCP), which runs a weighted version of TCP on each path so as to control the total aggregate bandwidth [22]. In EWTCP, a multipath flow can get the same throughput as a single TCP flow over the bottleneck link. However, EWTCP is shown to be inefficient in terms of network utilization [74]. The congestion control on each path is independent, so EWTCP cannot automatically switch the traffic to the less congested path. The authors of [33] propose a scalable TCP solution for multipath transmission, which adjusts the sending rate over each path based on the total congestion situation of all paths. The scalable TCP scheme is prone to switching a large portion of traffic to less congested paths, which may have a bandwidth smaller than that of other paths. As a result, the multipath transmission may not be able to provide a large aggregate throughput to mobile users.

2.3 Multipath TCP (MPTCP)

2.3.1 MPTCP Standard

Since 2009, IETF has worked on the multipath TCP (MPTCP) protocol which adds the capability of using multiple paths simultaneously to a regular TCP connection. So far, the IETF Multipath TCP group has already published five Request for Comments (RFC), which state different aspects of the MPTCP protocol, such as the architecture [17], congestion control algorithm [52], implementation guideline [16], APIs to the application layer [59] and security considerations [5].

The design of the MPTCP structure is based on the ideas of transport next-generation (TNG) [18], which summarizes many lessons learned from previous research and development practice for the transport layer. Specifically, TNG divides the transport layer into two sublayers, namely, the application-oriented layer and the network-oriented layer. On one hand, the application-oriented layer provides functions which support and protect the application's end-to-end communications. On the other hand, the network-oriented layer mainly focuses on the functions of endpoint identification (e.g., using port numbers in TCP) and congestion control. The above design principle offers a new perspective on extension of the Internet architecture and its bearing on the design of any new Internet transports or transport extensions [17].

As one of the TNG instantiations, as shown in Fig. 2.1, MPTCP loosely splits the transport layer into two sublayers; namely, MPTCP and subflow TCP. Here, MPTCP can be seen as the application-oriented layer while subflow TCP is the network-oriented layer. Based on this architecture, MPTCP can be easily implemented within the current network protocol stack. Subflow TCP runs on each path independently and reuses most functions of the regular TCP.

One of the differences between subflow TCP and regular TCP lies in that congestion control on each path is delegated to the MPTCP sublayer. The regular TCP uses congestion window to control the sending rate at the source. Although each subflow

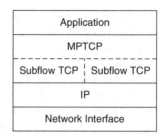

Application		Application	
		MPTCP	
TCP		Subflow TCP	Subflow TCP
IP		IP	
Network Interface		Network Interface	

Fig. 2.1 Comparison of protocol stacks with regular TCP and MPTCP

maintains a congestion window at the source (sender), the coupled congestion control algorithm [52] is designed so that MPTCP flow should (1) perform at least as well as a single-path flow would on the best of the paths available to it; and (2) take no more capacity than a single-path flow would obtain at maximum when experiencing the same loss rate. Basically, the requirement (1) motivates users to run MPTCP, while the requirement (2) guarantees that an MPTCP flow gracefully shares the path bandwidth with regular single-path TCP flows. Specifically, the MPTCP congestion control algorithm works as follows:

- Once the source receives an acknowledgement (ACK) from path r, it increases the congestion window of path r by $\min(a/w_{total}, 1/w_r)$; and
- Once the source receives a congestion signal from path r, it decreases the congestion window w_r of path r to $w_r/2$.

Here, w_r is the current congestion window size (in the unit of MSS) of subflow on path r, w_{total} is the total congestion window size of all subflows, given by $w_{total} = \sum_{r=1}^{K_s} w_r$, where K_s is the number of subflows, and a is an *aggressiveness* factor defined by

$$a = w_{total} \frac{\max_{1 \le r \le K_s} \frac{w_r}{RTT_r^2}}{\left(\sum_{r=1}^{K_s} \frac{w_r}{RTT_r}\right)^2}.$$ (2.1)

In (2.1), RTT_r is the RTT of path r. The increment $\min(a/w_{total}, 1/w_r)$ for congestion window size aims to ensure that each MPTCP subflow does not increase its congestion window faster than a single-path TCP flow with the same window size.

In addition to the aforementioned congestion control algorithm, another key component of MPTCP is packet reordering for multiple paths. As each subflow TCP maintains an independent sequence number space, the destination (receiver) may receive two packets of the same sequence number. Further, the packets received at the destination can be out-of-order [39] because of mismatched round-trip time (RTT) of multiple paths. Therefore, the source needs to tell the destination how to reassemble the data before delivering them to the application.

MPTCP solves this problem by using two levels of sequence numbers. First, the sequence number for subflow TCP is referred to as *subflow sequence number* (SSN), which is similar to the one in regular TCP. The subflow sequence number independently works within each subflow and ensures that data packets of each subflow are successfully transmitted to the destination in order. The sequence number at the MPTCP level is called *data sequence number* (DSN). Each packet received at the destination has a unique DSN no matter which path it is sent over. Hence, the destination can easily sequence and reassemble packets from different paths by DSN.

Moreover, the MPTCP sublayer is responsible for path management that discovers, adds and deletes subflows for the multipath connection between two hosts. Specifically, MPTCP supports such operations by defining new options in the MPTCP header [16]. For instance, the *Add Address* (ADD_ADDR) option announces additional addresses (and optionally, ports) on which a host can be reached. An ADD_ADDR option can be sent on an existing subflow, informing the receiver of the sender's alternative address(es). The recipient can use this information to open a new subflow to the sender's additional address. On the other hand, if, during the lifetime of an MPTCP connection, a previously announced address becomes invalid, the affected host should announce this through the *Remove Address* (REMOVE_ADDR) option so that the peer can terminate any subflows currently using that address.

2.3.2 MPTCP Extensions

This section reviews existing studies on MPTCP and extensions to MPTCP in the literature, which involve several important aspects of multipath transmission, such as achievable performance, mobility support, security, fairness and goodput.

There have been many studies on the MPTCP performance in different wireless network environments. The authors of [55] evaluate the performance of MPTCP in a data center network (e.g., Amazon EC2). The results show that MPTCP can effectively and seamlessly utilize the available bandwidth and achieve a fairness goal with various network topologies. The authors of [49, 53] implement MPTCP to support a make-before-break handover between 3G and Wi-Fi links in an opportunistic mobility scenario. They argue that the best level that handles mobility is the transport layer.

There are also some proposed mechanisms that extend the existing functionalities of MPTCP to achieve specific objectives. Pluntke et al. propose a scheduler at the source to minimize the energy consumption [51]. They define an energy model for each radio interface and gather the communication history of a mobile user continuously. Thus, a broad range of applications can be supported by customizing the scheduler via solving a Markov decision process offline.

Security is another important issue of MPTCP. In [48], the MPTCP handshake procedure is extended to reuse keys negotiated by the application layer protocol above it, such as secure sockets layer (SSL)/transport layer security (TLS) to authenticate additional subflows. Diez et al. propose several solutions to protect MPTCP from flooding and hijacking attacks by using hash chains [14].

Although MPTCP can re-order the packets at the destination based on the data sequence number, it does not offer any solution to avoid the out-of-order packets, which can jeopardize the goodput for the application layer. Goodput is the actual effective throughput to the application, which is the amount of in-order data received per time unit. Starting from 2012, there are some studies on the goodput performance of MPTCP. In [54], the goodput of MPTCP is enhanced by appropriately selecting the path for packet retransmission. When the slow path is blocked by a full receive buffer due to too many out-of-order packets, the source will retransmit packets over the fast path. As this scheme is only triggered when the receive buffer is full, it cannot handle goodput degradation in normal transmission stages. The schemes in [40] and [13] utilize network coding to recover packet loss at the destination and in turn increase the goodput. In such coding-based schemes, the source transmits the original data in one subflow and linear combinations of original data in another subflow. As such, the redundant coded data are utilized to recover lost and delayed packets. These schemes, however, require the support of network coding in both communication peers.

In addition, the fairness of multipath transmission is an interesting area that has been explored in many studies. For example, there have been some studies on fair bandwidth sharing between multipath and single-path TCP flows. In [7], the authors extend the definition of fairness from single-path transmission to multipath transmission. They examine four congestion control approaches including MPTCP with respect to the fairness. A multipath congestion control mechanism, dynamic window coupling (DWC), is proposed in [20], which aims to achieve both fair sharing and throughput maximization. DWC exploits the correlation between the path loss and delay to detect a bottleneck link shared by multiple paths. As such, subflows on paths of a common bottleneck can be grouped for the same congestion control. Then, congestion windows across subflow-sets, each having a distinct bottleneck, are considered independent to maximize the aggregate throughput. The work in [20] focuses on a general wired network with selected bottleneck scenarios.

References

1. Al, A.A.E., Saadawi, T., Lee, M.: LS-SCTP: A bandwidth aggregation technique for stream control transmission protocol. Computer Communications **27**(10), 1012–1024 (2004)
2. Albiero, F., Katz, M., Fitzek, F.H.: Energy-efficient cooperative techniques for multimedia services over future wireless networks. In: Proc. IEEE ICC (2008)
3. Ananthanarayanan, G., Padmanabhan, V., Ravindranath, L., Thekkath, C.: COMBINE: Leveraging the power of wireless peers through collaborative downloading. In: Proc. ACM MOBISYS (2007)
4. Anghel, P.A., Kaveh, M.: Exact symbol error probability of a cooperative network in a Rayleigh-fading environment. IEEE Trans. Wireless Commun. **3**(5), 1416–1421 (2004)

5. Bagnulo, M.: Threat analysis for TCP extensions for multipath operation with multiple addresses. IETF RFC 6181 (2011)
6. Bao, X., Li, J.: Generalized adaptive network coded cooperation (GANCC): A unified framework for network coding and channel coding. IEEE Trans. Commun. **59**(11), 2934–2938 (2011)
7. Becke, M., Dreibholz, T., Adhari, H., Rathgeb, E.: On the fairness of transport protocols in a multi-path environment. In: Proc. IEEE ICC (2012)
8. Bletsas, A., Khisti, A., Reed, D., Lippman, A.: A simple cooperative diversity method based on network path selection. IEEE J. Select. Areas Commun. **24**(3), 659–672 (2006)
9. Bonaventure, O.: Apple seems to also believe in multipath TCP. http://perso.uclouvain.be/olivier.bonaventure/blog/html/2013/09/18/mptcp.html (2013)
10. Chakravorty, R., Agarwal, S., Banerjee, S., Pratt, I.: A mobile bazaar for wide-area wireless services. Wireless Networks **13**(6), 757–777 (2007)
11. Chen, D., Ji, H., Leung, V.: Distributed best-relay selection for improving TCP performance over cognitive radio networks: A cross-layer design approach. IEEE J. Select. Areas Commun. **30**(2), 315–322 (2012)
12. Cover, T.M., Gamal, A.A.E.: Capacity theorems for the relay channel. IEEE Trans. Info. Theory **25**(5), 572–584 (1979)
13. Cui, Y., Wang, X., Wang, H., Pan, G., Wang, Y.: FMTCP: A fountain code-based multipath transmission control protocol. In: Proc. IEEE ICDCS (2012)
14. Diez, J., Bagnulo, M., Valera, F., Vidal, I.: Security for multipath TCP: A constructive approach. Internetional Journal of Internet Protocol Technology **6**(3), 146–155 (2011)
15. Fareed, M.M., Uysal, M.: On relay selection for decode-and-forward relaying. IEEE Trans. Wireless Commun. **8**(7), 3341–3346 (2009)
16. Ford, A., Raiciu, C., Handley, M., Bonaventure, O.: TCP extensions for multipath operation with multiple addresses. IETF RFC 6824 (2013)
17. Ford, A., Raiciu, C., Handley, M., Barre, S., Iyengar, J.: Architectural guidelines for multipath TCP development. IETF RFC 6182 (2011)
18. Ford, B., Lyengar, J.: Breaking up the transport logjam. In: Proc. ACM HOTNETS (2008)
19. Gergely, T., Long, T., Janos, L.: Energy efficient reliable cooperative multipath routing in wireless sensor networks. World Academy of Science, Engineering and Technology (44), 1376–1381 (2010)
20. Hassayoun, S., Iyengar, J., Ros, D.: Dynamic window coupling for multipath congestion control. In: Proc. IEEE ICNP (2011)
21. He, L.: Efficient multi-path routing in wireless sensor networks. In: Proc. International Conference on Wireless Communications, Networking and Mobile Computing (2010)
22. Honda, M., Nishida, Y., Eggert, L., Sarolahti, P., Tokuda, H.: Multipath congestion control for shared bottleneck. In: Proc. PFLDNeT Workshop (2009)
23. Hsieh, H., Sivakumar, R.: pTCP: An end-to-end transport layer protocol for striped connections. In: Proc. IEEE ICNP (2002)
24. Hsieh, H., Sivakumar, R.: A transport layer approach for achieving aggregate bandwidths on multi-homed mobile hosts. Wireless Networks **11**(2), 99–114 (2005)
25. Hsu, C., Su, H., Lin, P.: Joint subcarrier pairing and power allocation for ofdm transmission with decode-and-forward relaying. IEEE Trans. Signal Processing **59**(1), 399–414 (2011)
26. Hu, Z., Li, G.: On energy-efficient TCP traffic over wireless cooperative relaying networks. EURASIP Journal Wireless Communications and Networking (2012)
27. Hunter, T.E., Nosratinia, A.: Cooperation diversity through coding (2002)
28. Hunter, T.E., Nosratinia, A.: Diversity through coded cooperation. IEEE Trans. Wireless Commun. **5**(2), 1536–1276 (2006)
29. Ibrahim, A.S., Sadek, A.K., Su, W.: Cooperative communications with relay-selection: When to cooperate and whom to cooperate with? IEEE Trans. Wireless Commun. **7**(7), 2814–2827 (2008)
30. Issariyakul, T., Krishnamurthy, V.: Amplify-and-forward cooperative diversity wireless networks: Model, analysis, and monotonicity properties. IEEE/ACM Trans. Networking **17**(1), 225–238 (2009)

31. Iyengar, J., Amer, P., Stewart, R.: Concurrent multipath transfer using SCTP multihoming over independent end-to-end paths. IEEE/ACM Trans. Networking **14**(5), 951–964 (2006)
32. Ju, P., Song, W., Zhou, D.: Survey on cooperative medium access control protocols. IET Communications **7**(9), 893–902 (2013)
33. Kelly, T.: Scalable TCP: Improving performance in highspeed wide area networks. ACM SIGCOMM Computer Communication Review **33**(2), 83–91 (2003)
34. Kim, K., Shin, K.: Improving TCP performance over wireless networks with collaborative multi-homed mobile hosts. In: Proc. ACM MOBISYS (2005)
35. Kramer, G., Gastpar, M., Gupta, P.: Cooperative strategies and capacity theorems for relay networks. IEEE Trans. Info. Theory **51**(9), 3037–3063 (2005)
36. Kuo, C., Wang, C., Lin, J.: Cooperative wireless broadcast for scalable video coding. IEEE Trans. Circuits Syst. Video Technol. **21**(6), 816–824 (2011)
37. Kwasinski, A.: Transmission of TCP traffic over user cooperative communications in infrastructure networks. In: Proc. IEEE WCNC (2010)
38. Laneman, J.N., Wornell, G.W., Tse, D.N.C.: An efficient protocol for realizing cooperative diversity in wireless networks. In: Proc. IEEE ISIT (2001)
39. Leung, K., Li, V., Yang, D.: An overview of packet reordering in transmission control protocol (TCP): Problems, solutions, and challenges. IEEE Trans. Parallel Distrib. Syst. **18**(4), 522–535 (2007)
40. Li, M., Lukyanenko, A., Cui, Y.: Network coding based multipath TCP. In: Proc. IEEE INFOCOM Computer Communication Workshop (2012)
41. Liao, I., Wang, J., Zhu, X.: cmpSCTP: An extension of SCTP to support concurrent multi-path transfer. In: Proc. IEEE ICC (2008)
42. Liu, P., Tao, Z., Narayanan, S., Korakis, T., Panwar, S.S.: CoopMAC: A cooperative MAC for wireless LANs. IEEE J. Select. Areas Commun. **25**(2), 340–354 (2007)
43. Liu, X., Cheung, G., Chuah, C.: Structured network coding and cooperative local peer-to-peer repair for MBMS video streaming. In: IEEE 10th Workshop on Multimedia Signal Processing (2008)
44. Liu, Y., Hefeeda, M.: Video streaming over cooperative wireless networks. In: Proc. IEEE ICC (2010)
45. Moualeu, J.M., Xu, H., Takawira, F.: Turbo codes in coded cooperation using the forced symbol method. In: Proc. IEEE WCNC (2009)
46. Mueller, S., Tsang, R., Ghosal, D.: Multipath routing in mobile ad hoc networks: Issues and challenges. Performance Tools and Applications to Networked Systems pp. 209–234 (2004)
47. Nosratinia, A., Hunter, T.E., Hedayat, A.: Cooperative communication in wireless networks. IEEE Commun. Mag. **42**(10), 74–80 (2004)
48. Paasch, C., Bonaventure, O.: Securing the multipath TCP handshake with external keys. IETF draft-paasch-mptcp-ssl-00 (2012)
49. Paasch, C., Detal, G., Duchene, F., Raiciu, C., Bonaventure, O.: Exploring mobile/WiFi handover with multipath TCP. In: Proc. ACM SIGCOMM Workshop on Cellular Networks: Operationis, Challenges, and Future Design (CellNet) (2012)
50. Pathmasuritharam, J., Das, A., Gupta, A.: Efficient multi-rate relaying (EMR) MAC protocol for ad hoc networks. In: Proc. IEEE ICC (2005)
51. Pluntke, C., Eggert, L., Kiukkonen, N.: Saving mobile device energy with multipath TCP. In: Proc. ACM MOBIARCH (2011)
52. Raiciu, C., Handley, M., Wischik, D.: Coupled congestion control for multipath transport protocols. IETF RFC 6356 (2011)
53. Raiciu, C., Niculescu, D., Bagnulo, M., Handley, M.: Opportunistic mobility with multipath TCP. In: Proc. ACM MOBIARCH (2011)
54. Raiciu, C., Paasch, C., Barre, S., Ford, A.: How hard can it be? Designing and implementing a deployable multipath TCP. In: Proc. USENIX NSDI (2012)
55. Raiciu, C., Barre, S., Pluntke, C., Greenhalgh, A., Wischik, D., Handley, M.: Improving datacenter performance and robustness with multipath TCP. In: Proc. ACM SIGCOMM (2011)
56. Ramadan, M., Zein, E., Dawy, Z.: Implementation and evaluation of cooperative video streaming for mobile devices. In: Proc. IEEE PIMRC (2008)

57. Rappaport, T.S.: Wireless Communications: Principles and Practice. Prentice Hall (2002)
58. Ribeiro, A., Cai, X., B.Giannakis, G.: Symbol error probabilities for general cooperative links. IEEE Trans. Wireless Commun. **4**(3), 1264–1273 (2006)
59. Scharf, M., Ford, A.: Multipath TCP (MPTCP) application interface considerations. IETF RFC 6897 (2013)
60. Schwarz, H., Marpe, D., Wiegand, T.: Overview of the scalable video coding extension of the H.264/AVC standard. IEEE Trans. Circuits Syst. Video Technol. **17**(9), 1103–1120 (2007)
61. Seenivasan, T.V., Claypool, M.: CStream: Neighborhood bandwidth aggregation for better video streaming. Multimedia Tools and Applications pp. 1–30 (2011)
62. Sendonairs, A., Erkip, E., Aazhang, B.: User cooperation diversity - Part I: System description. IEEE Trans. Commun. **51**(11), 1927–1938 (2003)
63. Sendonairs, A., Erkip, E., Aazhang, B.: User cooperation diversity - Part II: Implementation aspects and performance analysis. IEEE Trans. Commun. **51**(11), 1939–1948 (2003)
64. Shan, H., Cheng, H., Zhuang, W.: Cross-layer cooperative mac protocol in distributed wireless networks. IEEE Trans. Wireless Commun. **10**(8), 2603–2615 (2011)
65. Shan, H., Zhuang, W., Wang, Z.: Distributed cooperative mac for multihop wireless networks. IEEE Commun. Mag. **47**(2), 126–133 (2009)
66. Sharma, P., Lee, S., Brassil, J., Shin, K.G.: Handheld routers: Intelligent bandwidth aggregation for mobile collaborative communities. In: Proc. BroadNets (2004)
67. Sheng, Z., Ding, Z., Leung, K.: Cooperative communications in multi-hop wireless networks: Joint flow routing and relay node assignment. In: Proc. IEEE INFOCOM (2010)
68. Simoen, S., Munoz, O., Vidal, J.: Achievable rates of compress-and-forward cooperative relaying on gaussian vector channels. In: Proc. IEEE ICC (2007)
69. Simoens, S., Vidal, J., Munoz, O.: Compress-and-forward cooperative relaying in mimo-ofdm systems. In: Proc. IEEE SPAWC (2006)
70. Stewart, R., Xie, Q., Morneault, K., et al.: Stream control transmission protocol. IETF RFC 4960 (2007)
71. Tullimas, S., Nguyen, T., Edgecomb, R.: Multimedia streaming using multiple TCP connections. ACM Trans. Multimedia Computing, Communications and Appllications **4**(2) (2008)
72. Wei, Y., Song, M., Yu, F.: TCP performance improvement in wireless networks with cooperative communications and network coding. In: Proc. IEEE ICC (2012)
73. Wei, Y., Yu, F., Song, M., Zhang, Y.: Cross-layer design for TCP throughput optimization in cooperative relaying networks. In: Proc. IEEE ICC (2010)
74. Wischik, D., Raiciu, C., Greenhalgh, A., Handley, M.: Design, implementation and evaluation of congestion control for multipath TCP. In: Proc. USENIX NSDI (2011)
75. Xu, H., Huang, L., Qiao, C., Zhang, Y., Sun, Q.: Bandwidth-power aware cooperative multipath routing for wireless multimedia sensor networks. IEEE Trans. Wireless Commun. **11**(4), 1532–1543 (2012)
76. Yang, Z., Yao, Y., Li, X., Zheng, D.: A TDMA-based mac protocol with cooperative diversity. IEEE Commun. Lett. **14**(6), 542–544 (2010)
77. Zhang, M., Lai, J., Krishnamurthy, A., Peterson, L., Wang, R.: A transport layer approach for improving end-to-end performance and robustness using redundant paths. In: Proc. USENIX ATC (2004)
78. Zhao, G., Yang, C., Li, G., Li, D., Soong, A.: Channel allocation for cooperative relays in cognitive radio networks. In: Proc. IEEE ICASSP (2010)
79. Zhou, T., Sharif, H., Hempel, M., Mahasukhon, P., Wang, W., Ma, T.: A novel adaptive distributed cooperative relaying MAC protocol for vehicular networks. IEEE J. Select. Areas Commun. **29**(1), 72–82 (2011)
80. Zhu, H., Cao, G.: rDCF: A relay-enabled medium access control protocol for wireless ad hoc networks. IEEE Trans. Mobile Comput. **5**(9), 1201–1214 (2006)
81. Zhuang, W., Ismail, M.: Cooperation in wireless communication networks. IEEE Wireless Commun. Mag. **19**(2), 10–20 (2012)
82. Zou, S., Li, B., Wu, H., Zhang, Q., Zhu, W., Cheng, S.: A relay-aided media access (RAMA) protocol in multirate wireless networks. IEEE Trans. Veh. Technol. **55**(5), 1657–1667 (2006)

Chapter 3
System Modelling

This chapter first presents the system setup for the user cooperation scenario that is considered in this brief for the LTE network. Then, it introduces the traffic model and the default parameters to test such a system, and the framework of the system, including different function modules.

3.1 LTE Network with User Cooperation

This brief presents state-of-the-art extensions to the MPTCP protocol [2] for the user cooperation scenario so as to offer a stable QoS to mobile users. Figure 3.1 shows a user cooperation scenario with three user equipment (UE) devices associated with an Evolved Node B (eNB). In this case, UE 3 is the destination that receives data from the application server (source). Two nearby relays (UE 1 and UE 2) can receive packets on behalf of the destination (UE 3) via their own LTE links and then forward the packets toward the destination via Wi-Fi links. Hence, the destination can aggregate the available bandwidth of the two relays.

In such a user cooperation scenario, MPTCP can run multiple TCP subflows between the source and the destination through multiple relays. Since only one IP address is usually allocated to one network interface card (NIC), there is a potential problem when the destination is required to have multiple IP addresses for its Wi-Fi interface to establish multiple subflow connections. Fortunately, this can be solved by the virtual interface technique to configure multiple IP addresses to one NIC [6].

Moreover, it is assumed that each subflow TCP runs over an independent path in the wired network between the application server and the eNB. Hence, there are no shared bottleneck links in the wired network. Additionally, because the Wi-Fi links usually have a higher transmission rate than the LTE links between the eNB and the relays, the LTE links are assumed to be the bottleneck links of the end-to-end paths between the source and the destination.

© The Author(s) 2014
D. Zhou, W. Song, *Multipath TCP for User Cooperation in Wireless Networks*,
SpringerBriefs in Computer Science, DOI 10.1007/978-3-319-11701-0_3

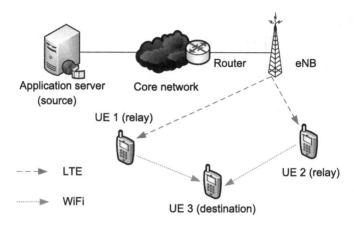

Fig. 3.1 User cooperation in the LTE network

Table 3.1 Default system parameters

Parameter	Value
Transmit power of eNB	30 dBm
Transmit power of UE	23 dBm
Noise figure at eNB	5 dB
Noise figure at UE	5 dB
Transmission time interval (TTI)	1 ms
eNB scheduler	Blind equal throughput (BET)
Radio link control (RLC) mode	Acknowledge mode (AM)
Adaptive modulation & coding (AMC)	PiroEW2010 [4]
Number of resource blocks (RBs)	50
Fading channel trace	Pedestrian at 3 km/h
Wi-Fi link	IEEE 802.11a
Wi-Fi transmission rate	54 Mbit/s
Wi-Fi fragmentation threshold	2,200 bytes
Wi-Fi RTS/CTS threshold	2,200 bytes
Number of available relays	10

The throughput of each MPTCP subflow is then dependent on how much available bandwidth a relay can provide to the destination. In the LTE network, one key factor that impacts the available bandwidth is the scheduler used at the eNB. A different scheduler can allocate a different amount of resources to a relay. The relay then offers a different available bandwidth to the destination. In order to exclude the influence of the LTE scheduler on the performance of the proposed extension modules, the blind equal throughput (BET) scheduler [1] is considered at the eNB, which aims to provide an equal throughput to all UEs associated with the eNB. The default system parameters used in the experiments of this brief for the user cooperation scenario are given in Table 3.1.

3.2   Traffic Model for Multipath Transmission

In order to exclude the impact of different traffic patterns of various applications, the bulk data traffic is considered for multipath flow between the source and the destination. Specifically, the traffic generator at the source ensures that there are always data to transmit and the source sends over the data as fast as possible. Once the sending buffer of the source is filled, the generator suspends and the source resumes sending out the data when enough space is cleared in the sending buffer, e.g., to accommodate at least one packet.

Three types of single-path flows, based on UDP, TCP and AIMD, are considered between the source and each relay to simulate the background traffic running at the relays. These single-path flows can follow two types of traffic patterns: (1) a static traffic pattern in which the sending rate does not change during the connection; and (2) a dynamic traffic pattern in which the sending rate is varying during the connection. In order to implement such traffic patterns, an on-off traffic generator can be used to control the sending rate of a single-path flow. As shown in Fig. 3.2, the on-off traffic generator alternates between the ON and OFF states. During the ON state, the traffic generator produces data of a constant bit rate (CBR). During the OFF state, the traffic generator just suspends. The durations of ON and OFF states, referred to as on time and off time, respectively, are random and follow two exponential distributions. As such, the dynamic traffic pattern can be simulated by setting different data rates for the on time. In contrast, the static traffic pattern can be implemented by using a constant data rate for the entire connection time without involving the on-off behavior.

3.3   Structure of Multipath Enhancement Modules

For the system illustrated in Fig. 3.1, the source, the relays and the destination are all assumed to be MPTCP-enabled nodes. Moreover, each node can be further augmented with the extension modules that will be introduced in this brief to enhance the multipath transmission performance. The basic ideas of these modules are introduced in the following, together with their positions in the protocol stacks of the source, the relays and the destination, which are shown in Fig. 3.3.

First, the subset-sum based relay selection (SSRS) module proposed in [9] is located at the relay and the destination. It aims to guarantee a stable aggregate throughput that satisfies a target bit rate (TBR) requirement of the application layer

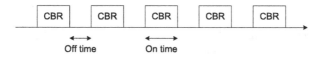

Fig. 3.2 On-off traffic model

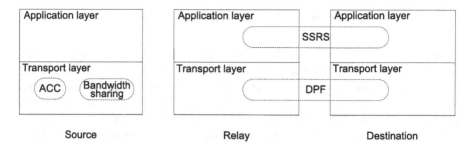

Fig. 3.3 System framework

at the destination. The key idea is that the destination maintains multiple relay sets whose total available bandwidths are within an acceptable TBR range (e.g., between 90 and 110 % of TBR). Once the total available bandwidth of the relay set is detected out of this range, the destination updates the current in-use relay set to a new set so as to maintain a stable aggregate throughput.

Second, the adaptive congestion control (ACC) module proposed in [10] can further improve the goodput at the destination by extending the coupled congestion control algorithm of MPTCP and achieving similar end-to-end delays over multiple paths. The main idea is to have the source dynamically adjust the congestion window of each subflow TCP so as to relieve the traffic load on a slow path and reduce the corresponding end-to-end path delay. Many factors contribute to the end-to-end delay (e.g., transmission, processing, and queueing delays at routers, and packet retransmission over the wireless link), so ACC cannot eliminate all delay gaps among different paths. Therefore, the differentiated packet forwarding (DPF) module proposed in [7] can further complement ACC. In DPF, the destination informs each relay of the expected range of MPTCP data sequence numbers (DSNs). Thus, the relays can temporarily buffer the packets of a DSN outside the range and only forward the packets within the DSN range to the destination. The packets buffered at the relays will only be forwarded to the destination when the DSN range is updated by the destination and these buffered packets fall into the new range.

Third, in order to respect the local traffic at the relays, two bandwidth sharing schemes proposed in [11] are introduced to extend the MPTCP congestion control (MCC) algorithm. In the user cooperation scenario, if the throughput of local single-path flows at the relays is degraded by forwarding the MPTCP traffic for the destination, the relays will not be motivated to provide the relaying service. In addition, if the available bandwidth provided by the relays varies with the bandwidth competition, the performance of SSRS will also be affected. Hence, the first extension, referred to as *MCC-Coop*, can ensure that the MPTCP flow of the destination runs fairly with the local single-path TCP flows of the relays. To further protect local AIMD flows, another more generic extension, referred to as *GMCC-Coop*, is also proposed in [11]. Although both MCC-Coop/GMCC-Coop and ACC are extensions to MPTCP congestion control, they will not conflict with each other because they focus on different aspects of congestion control.

In an overall perspective, SSRS, MCC-Coop and GMCC-Coop can be seen as the foundation to ACC and DPF. As such, various types of the local traffic at the relays are protected, while a stable aggregate throughput is guaranteed for the MPTCP flow. Since the aggregate throughput can be viewed as the maximum goodput that the MPTCP flow can achieve, ACC and DPF further improve the goodput to approach the maximum limit. They can work independently and are also mutually complementary to each other.

In order to evaluate the performance of the above modules, a latest network simulator—ns-3 [3] can be used to conduct simulations in a variety of scenarios. Specifically, the MPTCP framework can be implemented in ns-3, which includes the core functions of MPTCP, such as socket APIs [5], coupled congestion control, path management, and packet scheduler. In addition, the LTE implementation in ns-3 need to be extended so as to construct the user cooperation scenario [8]. One extension of the LTE module can be found in the official release 3.16 of ns-3, which includes a new packet scheduler at the eNB and a new UE registration process.

References

1. Capozzi, F., Piro, G., Grieco, L., Boggia, G., Camarda, P.: Downlink packet scheduling in LTE cellular networks: Key design issues and a survey. IEEE Communications Surveys & Tutorials **15**(2), 678–700 (2012)
2. Ford, A., Raiciu, C., Handley, M., Barre, S., Iyengar, J.: Architectural guidelines for multipath TCP development. IETF RFC 6182 (2011)
3. NS-3: The network simulator - ns-3. http://www.nsnam.org/ (2013)
4. Piro, G., Grieco, L., Boggia, G., Camarda, P.: A two-level scheduling algorithm for QoS support in the downlink of LTE cellular networks. In: Proc. European Wireless Conference (EW) (2010)
5. Scharf, M., Ford, A.: Multipath TCP (MPTCP) application interface considerations. IETF RFC 6897 (2013)
6. Winter, R., Faath, M., Ripke, A.: Multipath TCP support for single-homed end-systems: draft-wr-mptcp-single-homed-05. Ietf internet-draft (2013)
7. Zhou, D., Ju, P., Song, W.: Performance enhancement of multipath TCP with cooperative relays in a collaborative community. In: Proc. IEEE PIMRC (2012)
8. Zhou, D., Song, W., Baldo, N., Miozzo, M.: Evaluation of TCP performance with lte downlink schedulers in a vehicular environment. In: Proc. IWCMC (2013)
9. Zhou, D., Song, W., Ju, P.: Subset-sum based relay selection for multipath TCP in cooperative LTE networks. In: Proc. IEEE GLOBECOM (2013)
10. Zhou, D., Song, W., Shi, M.: Goodput improvement for multipath TCP by congestion window adaptation in multi-radio devices. In: Proc. IEEE CCNC (2013)
11. Zhou, D., Song, W., Wang, P., Zhuang, W.: Multipath TCP for user cooperation in LTE networks. IEEE Network. http://cs.unb.ca/wsong/publications/journals/NET_MPTCP_Dizhi.pdf (2014)

Chapter 4
Subset-Sum Based Relay Selection (SSRS)

In a user cooperation scenario, the available bandwidth provided by relays can be highly varying due to a range of factors such as wireless channel fading, dynamic local traffic load at relays, and even the type of packet scheduler at the base station. As a result, it is challenging to maintain a stable aggregate throughput with MPTCP [2] over relays. This chapter introduces an enhancement module at the application layer for a user cooperation scenario in the LTE network. Based on the relay bandwidth monitoring, a subset-sum based relay selection (SSRS) module is developed in [9] for adding and deleting paths so as to ensure a stable aggregate throughput in a highly varying environment. The relay selection algorithm is based on a fully polynomial-time subset-sum approximation [1]. Extensive simulations are conducted to evaluate the SSRS module in different background traffic patterns. The simulation results well demonstrate the strengths of SSRS in minimizing throughput outage, the number of active subflows, and performance variation.

4.1 SSRS Module[1]

4.1.1 Structure of SSRS Module

In the user cooperation scenario shown in Fig. 3.1, due to the fading effect and the various local traffic load of relays, the available bandwidths offered by relays are highly dynamic, which can seriously impact the performance of bandwidth-intensive applications, e.g., video streaming. In order to guarantee a stable aggregate throughput of MPTCP in the user cooperation scenario, this section introduces a

---

[1]Reprinted with permission, from Proceedings of IEEE GLOBECOM 2013, "Subset-sum based relay selection for multipath TCP in cooperative LTE networks," by D. Zhou, W. Song, and P. Ju [9].

D. Zhou, W. Song, *Multipath TCP for User Cooperation in Wireless Networks*,
SpringerBriefs in Computer Science, DOI 10.1007/978-3-319-11701-0_4

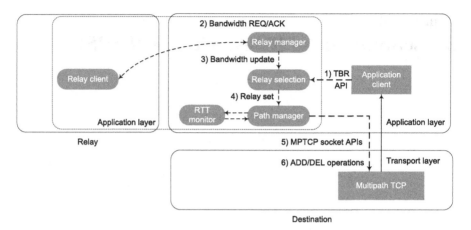

Fig. 4.1 Structure of SSRS

subset-sum based relay selection (SSRS) module [9], which consists of four main components, namely, the external application programming interfaces (API), relay manager, relay selection and path manager.

In SSRS, the relay manager is responsible for gathering the latest available bandwidth of relays periodically. After eliminating the relays with a high local traffic load, the relay manager forwards the selected relay list to the relay selection component, which further selects the feasible relay sets based on the available bandwidths and a target bit rate (TBR) configured by an application via the external API. Moreover, the relay selection component selects a best relay set from the feasible relay sets based on certain criteria. Then, the best relay set is forwarded to the path manager component, which monitors the aggregate throughput in a certain frequency. Whenever the aggregate throughput is observed out of a certain variation range of the TBR, the path manager directs the MPTCP component to update the current active relay set to the latest best relay set via MPTCP socket APIs. In practice, the path manager can obtain the aggregate throughput by registering a hook between the transport layer and the network layer so that all the incoming packets at the destination are counted by the path manager.

Figure 4.1 shows the structure of the SSRS module at the destination and the relay. The rectangular shapes show the existing components in the corresponding protocol stacks, while the rounded rectangular shapes in the dashed frame represent the components of the SSRS enhancement module. The SSRS module at the application layer has two external APIs with the standard protocol stacks. First, the SSRS module allows an application at the application layer to provide a desired TBR requirement. As such, different applications can customize their QoS requirements by setting different TBRs. For instance, TBRs ranging from 384 to 768 kbit/s are usually needed for the two-way interactive standard definition (SD) video conferencing, while the two-way high definition (HD) video conferencing requires a higher bandwidth support, which ranges from 768 kbit/s to 1.24 Mbit/s [7].

Meanwhile, by upper bounding the TBR, the source is prevented from occupying too much bandwidth by selecting the relays greedily, which guarantees the fairness. Second, SSRS uses the standard MPTCP socket APIs to enable multipath transmission [6]. In particular, `TCP_MULTIPATH_ADD` and `TCP_MULTIPATH_REMOVE` APIs are used to add and delete subflow paths, respectively. As such, SSRS can be easily deployed in MPTCP-enabled nodes without modifying the existing protocol stacks.

4.1.2 Operation Procedures

Based on the structure of SSRS given in Fig. 4.1, the SSRS module assists with the multipath transmission according to the following operation procedures.

At the beginning of an application session, the relay selection component in SSRS acquires the desired TBR from the application client, and the relay manager retrieves the available bandwidths of the relays from their periodic broadcasts at a frequency F, e.g., one broadcast per 2 s. The available bandwidth is obtained by the relay client by measuring its average local traffic load.

Once the relay manager obtains the available bandwidths of the relays, it forwards such information to the relay selection component, which further determines all feasible relay sets whose total available bandwidths are within an acceptable TBR range, e.g., between 90 and 110 % of TBR. Moreover, the relay selection component selects a best relay set based on certain criteria and configures the set as the active set. The algorithm of determining feasible relay sets and selecting the best relay set is discussed in Sect. 4.1.3. The active set is forwarded to the path manager at the application layer, which further calls MPTCP socket APIs to add all subflows to the MPTCP connection.

During an application session, since the relays periodically broadcast their available bandwidths to the relay manager, the relay selection component needs to update the feasible relay sets accordingly, so that their total available bandwidths can satisfy the acceptable TBR range. A new best relay set is selected as the backup set for the current active set. Once the total available bandwidth of the active set is found to be outside the TBR range, the path manager is triggered to migrate the current active set to the backup set. The detailed algorithm of selecting the backup set is also discussed in Sect. 4.1.3.

To migrate to a new relay set, the path manager compares the active set and the backup set to derive the required operations of adding and/or deletion subflows. Specifically, the new subflows are added first, whereas the deletion of the old subflows starts after a maximum RTT of the active paths. The main consideration of postponing deleting operations is to ensure that the destination waits to receive the packets on the fly over the paths to be deleted. Here, an RTT monitor is required to measure the RTT of each active path and provide such information to the path manager.

4.1.3   Relay Set Selection Algorithm

As discussed in Sect. 4.1.2, a real-time fast algorithm is necessary to efficiently select and update the feasible relay sets whose total available bandwidths satisfy the TBR range. This is the well-known subset-sum problem, which is proved to be NP-complete [1]. Given a set of N elements, there are totally 2^N possible subsets so that the searching scale is exponential.

Fortunately, a fully polynomial-time approximation algorithm is available to "trim" subsets that have sums sufficiently close to neighboring subsets [1]. This approximation algorithm is first adapted to obtain the feasible relay sets. The original approximation algorithm can determine the relay subsets whose total available bandwidths add up to an exact given value. Here, the relay selection algorithm needs to find relay subsets whose total available bandwidths fall into a range $[(1 - \theta)\text{TBR}, (1 + \theta)\text{TBR}]$, $0 < \theta < 1$. This is because the available bandwidth of each relay path may vary dynamically and a small buffer space is necessary to tolerate a certain level of throughput variation.

Given N relays, let b_i to denote the available bandwidth of relay i $(1 \le i \le N)$ and define $R = \{b_1, b_2, \ldots, b_N\}$. All possible total available bandwidths of relays $\{n_1, n_2, \ldots, n_i\}$ are denoted by

$$L_i = \{S_1^i, S_2^i, \ldots, S_{|L_i|}^i\} \tag{4.1}$$

$$(1 - \theta)\text{TBR} \le S_j^i \le (1 + \theta)\text{TBR}, \quad 1 \le j \le |L_i|.$$

All subsets of relays that have a total available bandwidth S_j^i are denoted by $\mathbb{U}_j^i(S_j^i)$. That is, $\forall X \in \mathbb{U}_j^i(S_j^i)$, X satisfies

$$\sum_{n_k \in X} b_k = S_j^i. \tag{4.2}$$

Algorithm 1 shows the iterative procedure to obtain the feasible relay subsets. The input parameters of Algorithm 1 include a set R of available bandwidth of each relay, an approximation parameter ε and the number N of relays. The outputs of Algorithm 1 include the set of total available bandwidths L_f, which are in the range of TBR, and the corresponding relay sets $\mathbb{U}_j^N(S_j^N)$ whose total available bandwidths are given by the elements in L_f. Then, the best relay will be selected based on these relay sets in $\mathbb{U}_j^N(S_j^N)$.

In each round, from line 6 to line 11, a new bandwidth set L_i is calculated by combining the previous available bandwidth set L_{i-1} with a new set L_i', which is defined by adding the available bandwidth b_i of a new relay i to each element of L_{i-1}. As given in Line 6 of Algorithm 1, $L_i' = \{\hat{S}_1^i, \hat{S}_2^i, \ldots, \hat{S}_{|L_{i-1}|}^i\}$, where $\hat{S}_j^i = S_j^{i-1} + b_i, \forall 1 \le j \le |L_{i-1}|$. That means, L_i' lists the total available bandwidths of subsets of relays $\{n_1, n_2, \ldots, n_i\}$, and these subsets must include relay n_i. Meanwhile, from line 7 to line 9, the corresponding subsets of relays

---

Algorithm 1 Subset-sum based relay selection

Input: R, ε, N
Output: L_f and $\mathbb{U}_j^N(S_j^N)$
1: $L_f = \varnothing$
2: $L_0 = \{0\}$
3: $S_1^0 = \{0\}$
4: $\mathbb{U}_1^0 = \varnothing$
5: **for** $i = 1$ to N **do**
        // Consider a new relay i of available bandwidth b_i
6:     $L_i' = \{\hat{S}_1^i, \hat{S}_2^i, \ldots, \hat{S}_{|L_{i-1}|}^i\}$, where $\hat{S}_j^i = S_j^{i-1} + b_i, \forall 1 \le j \le |L_{i-1}|$
        // Add a new relay i into correspondent relay sets
7:     **for** $j = 1$ to $|L_i|$ **do**
8:         $\mathbb{U}_j^i(S_j^i) \leftarrow \{\hat{X} | \hat{X} = X \cup n_i, X \in \mathbb{U}_j^{i-1}(S_j^{i-1})\}$
9:     **end for**
        // Merge sets L_{i-1} and L_i'
        // Sort the combined set in descending order
10:    $L_i = \text{MergeSort}(L_{i-1}, L_i')$
11:    $L_i = \{S_1^i, S_2^i, \ldots, S_{|L_i|}^i\}$
        // Remove all elements of L_i that are greater than the TBR upper bound
12:    **for** $j = 1$ to $|L_i|$ **do**
13:        **if** $\forall S_j^i \in L_i, S_j^i > (1 + \theta)\text{TBR}$ **then**
14:            Remove S_j^i from L_i
15:        **else if** $(1 - \theta)\text{TBR} \le S_j^i \le (1 + \theta)\text{TBR}$ **then**
16:            **if** $S_j^i \notin L_f$ **then**
17:                $L_f \leftarrow L_f \cup S_j^i$
18:            **end if**
19:        **else**
20:            break
21:        **end if**
22:    **end for**
23:    $\text{Trim}(L_i, \varepsilon/(2N))$
24: **end for**
25: Return L_f and $\mathbb{U}_j^N(S_j^N)$ for all $1 \le j \le |L_f|$

$\mathbb{U}_j^i(S_j^i)$ are also updated by adding the new relay i to each subset. Next, the algorithm merges the previous set L_{i-1} and the new set L_i', and then sorts the combined set in a descending order. All elements of L_i greater than the TBR upper bound are removed because they are definitely greater than the TBR upper bound in the next round.

All elements of L_i that fall into the TBR range are further trimmed by introducing an approximation parameter ε ($0 < \varepsilon < 1$). Given two neighboring elements S_j^i and S_{j+1}^i, if they are sufficiently close to satisfy the following equation,

$$\frac{S_{j+1}^i}{1 + \frac{\varepsilon}{2N}} \le S_j^i \le S_{j+1}^i \tag{4.3}$$

then S_{j+1}^i is removed from L_i. Actually, ε is an indicator of the variance of the approximation result from the optimal solution. Since the user requires to ensure an approximate total bandwidth in the range $[(1 - \theta)\text{TBR}, (1 + \theta)\text{TBR}]$, it is natural to set $\varepsilon = \theta$. Compared with the original subset-sum algorithm in [1], Algorithm 1 does not increase the search space, so that the running time remains polynomial in both $1/\varepsilon$ and N.

Based on the feasible relay subsets obtained from Algorithm 1. Given K relay subsets, a best relay set can be selected according to the following criteria. Specifically, the active set and the backup set are selected based on two main factors, i.e., the difference between TBR and the total available bandwidth of a relay subset k (denoted by ΔB_k), and the number of relays (denoted by N_s^k). The two factors of relay subset k are normalized according to the following equations:

$$\alpha_B^k = \frac{\Delta B_k - \min_{1 \leq l \leq K} \Delta B_l}{\max_{1 \leq l \leq K} \Delta B_l - \min_{1 \leq l \leq K} \Delta B_j}, \quad \alpha_N^k = \frac{N_s^k - \min_{1 \leq l \leq K} N_s^l}{\max_{1 \leq l \leq K} N_s^l - \min_{1 \leq l \leq K} N_s^l}. \tag{4.4}$$

Then, the following priority index γ_k of relay subset k is defined as

$$\gamma_k = \frac{1}{\alpha_B^k + \alpha_N^k}. \tag{4.5}$$

The relay subset of the highest priority index is selected as the active set.

During the MPTCP connection, multiple paths can be established through the relays in the active set from the source to the destination. When the destination observes the total available bandwidth of the current active set out of the TBR range, the destination is triggered to switch to the backup set. The selection of the backup set further takes into account the number of path maintenance operations (denoted by $M_{ops}^{k_1,k_2}$), which is the number of required operations of adding and deleting paths to migrate from the active set k_1 to the backup set k_2. For example, if the active set includes relays $\{n_1, n_2, n_3\}$, the number of path maintenance operations for switching to the backup set $\{n_2, n_3, n_4\}$ will be 2, including one addition of a new path through n_4 and one deletion of an old path through n_1. The rationale behind is to minimize the transition period and ensure the smooth migration with a minor throughput variation. Similar to the factors ΔB_k and N_s^k, $M_{ops}^{k_1,k_2}$ is also normalized as follows:

$$\alpha_M^{k_1,k_2} = \frac{M_{ops}^{k_1,k_2} - \min_{\substack{1 \leq l \leq K \\ l \neq k_1}} M_{ops}^{k_1,l}}{\max_{\substack{1 \leq l \leq K \\ l \neq k_1}} M_{ops}^{k_1,l} - \min_{\substack{1 \leq l \leq K \\ l \neq k_1}} M_{ops}^{k_1,l}}. \tag{4.6}$$

Fig. 4.2 UE distribution

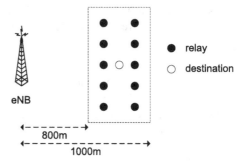

The priority index in (4.5) is then extended to

$$\tilde{\gamma}_{k_1,k_2} = \frac{1}{\alpha_B^{k_2} + \alpha_N^{k_2} + \alpha_M^{k_1,k_2}}. \tag{4.7}$$

The relay subset k_2 with the highest priority index $\tilde{\gamma}_{k_1,k_2}$ is selected as the backup set for the active set k_1.

4.2   Experimental Results of SSRS and Greedy[2]

To evaluate the performance of the SSRS module, SSRS can be implemented in the network simulator ns-3 [5] based on the system model introduced in Chap. 3. The simulation considers an eNB connected to 11 UEs, among which there are one destination and ten relays. These UEs are uniformly distributed within a rectangle area of a distance 800–1,000 m to the eNB as illustrated in Fig. 4.2. Equipped with both LTE and Wi-Fi interfaces, the relays and the destination can use their Wi-Fi interfaces to directly communicate in an ad hoc mode. The destination receives packets from the application server (the source) via the relays. The relay manager at the destination monitors the available bandwidths of the relays every 2 s. In other words, the frequency F equals 1 measurement per 2 s. The TBR at the destination is set to 3 Mbit/s, which satisfies the requirement of most video streaming service providers, such as YouTube© [8], Netflix© [4] and Hulu© [3]. The other default simulation parameters are given in Table 3.1.

The simulation evaluates the aggregate throughput and goodput achieved at the destination, and the number of subflows engaged in the MPTCP connection to verify the performance of SSRS. To simulate varying available bandwidths of the relays,

---

[2]Reprinted with permission from IET Communications (2014), "Goodput improvement for multipath transport control protocol in cooperative relay-based wireless networks," by D. Zhou, W. Song, and P. Ju [10].

the background traffic load based on UDP at the relays is adjusted. Specifically, two patterns are considered to evaluate the adaptiveness and effectiveness of SSRS:

- *Static available bandwidth pattern*: the UDP traffic rates do not change during the simulation time, so that only the wireless channel effects, such as fading, affect the available bandwidths via relays.
- *Dynamic available bandwidth pattern*: the UDP traffic rates at relays are increasing or decreasing linearly over the simulation time, so that the available bandwidths of the relays are decreased or increased correspondingly.

In addition, to better understand the performance of SSRS, a greedy relay selection scheme is considered as a benchmark, referred to as *Greedy* in the following figures. As shown in Algorithm 2, the relays are first sorted in an ascending order according to their available bandwidths. Then, a relay is added into the resulting relay set $\mathbb{V}$ one by one from the relay of the lowest available bandwidth, until the total available bandwidth falls into the acceptable TBR range. During the run time, if the total available bandwidth becomes greater than the TBR upper bound, Greedy deletes a relay one by one, also starting with the relay of the lowest available bandwidth, until the total available bandwidth returns to the TBR range.

Algorithm 2 Greedy relay selection

Input: R, N
Output: $\mathbb{V}$
 1: $\hat{R} = \text{Sort}(R)$
    // Sort relays in an ascending order of available bandwidth
 2: $S_0 = 0$
 3: $\mathbb{V} = \varnothing$
 4: **for** $i = 1$ to N **do**
 5:     **if** $S_{i-1} + b_i < (1 - \theta)TBR$ **then**
 6:         $\mathbb{V} \leftarrow \mathbb{V} \cup n_i$
 7:         $S_i = S_{i-1} + b_i$
 8:     **end if**
 9:     **if** $S_i \geq (1 + \theta)TBR$ **then**
10:         break
11:     **end if**
12: **end for**
13: Return $\mathbb{V}$

4.2.1   Static Scenario

Figure 4.3 compares the aggregate throughput and goodput of SSRS and Greedy in the static available bandwidth pattern. Two straight lines show the upper bound and lower bound of the TBR requirement, which are 3.3 Mbit/s and 2.7 Mbit/s,

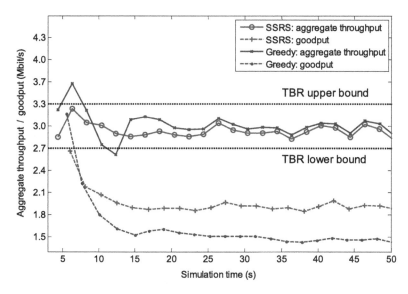

Fig. 4.3 Aggregate throughput and goodput of SSRS in the static available bandwidth pattern at relays

respectively. In the static scenario, only the channel fading affects the available bandwidth of each relay. As seen in Fig. 4.3, both SSRS and Greedy can guarantee the stable aggregate throughput in the long term. After the 15th s, both algorithms achieve a throughput around TBR with a variation less than 5 %.

However, SSRS has less TBR variation (6–8 %) than Greedy (13–23 %) in the beginning. This is because that SSRS has a larger search space for relay subsets than Greedy. It efficiently scans most feasible subsets of relays so as to select a relay set that provides a total bandwidth closest to TBR. In other words, SSRS can tolerate a larger variation because the total bandwidths of selected relay subsets are more concentrated in the middle of the TBR range. Therefore, there is a lower outage probability in the next update period.

In contrast, Greedy adds and deletes paths based on the current active set. As a result, there is a larger chance to select a relay set having a total available bandwidth close to the TBR edges. For example, at the 4th s, SSRS selects a backup set of a total available bandwidth 2.85 Mbit/s, whereas the backup set selected by Greedy has a total bandwidth 3.25 Mbit/s, which almost approaches the TBR upper bound 3.3 Mbit/s. At the 8th s, Greedy eventually violates the TBR boundary.

Figure 4.3 also shows that the goodput of SSRS and Greedy are both much lower than their aggregate throughput. The average goodput of SSRS is only 70 % of its aggregate throughput, while this ratio is 53 % for Greedy. One main reason for this is the number of subflows in use, which is shown in Fig. 4.4. When the number of subflows is growing, the end-to-end path delay variation is also increased so that more out-of-order packets are received at the destination. SSRS employs 1 less subflow than Greedy, so the average goodput of SSRS is 1.25 times that of Greedy.

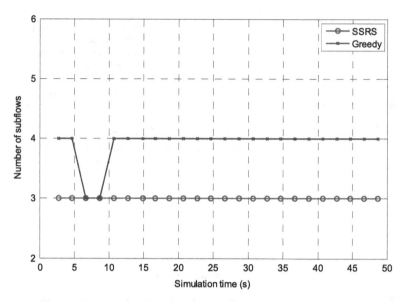

Fig. 4.4 The number of subflows of SSRS in the static available bandwidth pattern at relays

The reason for SSRS to choose less subflows is because of the priority index it uses for relay selection. As seen in Eq. (4.5), the relay set k that has a smaller number of subflows α_N^k ends up with a higher priority index γ_k. Therefore, even though the destination may switch relays during the simulation due to channel fading, the number of subflows is very stable for SSRS.

4.2.2   Dynamic Scenario

Figures 4.5 and 4.6 show the aggregate throughput, goodput and the number of subflows of SSRS and Greedy with the dynamic available bandwidth patterns. In Fig. 4.5a, the UDP traffic load at each relay is linearly increased by 0.2 Mbit/s at the 7th, 13rd and 19th s to simulate decreasing available bandwidths. Then, in Fig. 4.6b, the local UDP throughput at each relay is linearly decreased by 0.2 Mbit/s at the 31st, 37th and 47th s to simulate increasing available bandwidths.

As seen in Fig. 4.6a, although the aggregate throughput achieved at the destination goes down because of less available bandwidth at each relay, SSRS adapts more smoothly with less throughput variation and outage than Greedy. Among the ten monitor points, SSRS only has 1 outage, as opposed to 3 for Greedy. This is because SSRS effectively selects the backup set whose total available bandwidth is much closer to TBR. Thus, there is a larger guard space for throughput variation so as to minimize the possibility of throughput outage. Even so, both algorithms still need to employ more relays to accommodate the lower available bandwidth of

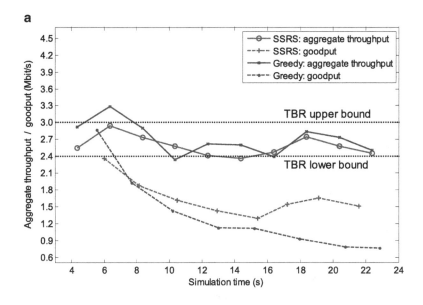

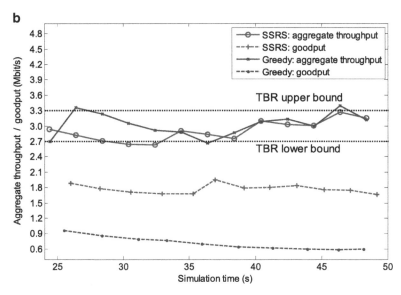

Fig. 4.5 Aggregate throughput and goodput of SSRS with dynamic available bandwidth patterns at relays.(**a**) Decreasing available bandwidth at relays. (**b**) Increasing available bandwidth at relays

each relay. Similar behavior is observed in Fig. 4.5b when the aggregate throughput achieved at the destination goes up because of more available bandwidth at each relay. During this period, Greedy suffers from 3 outage events while SSRS only has 1.

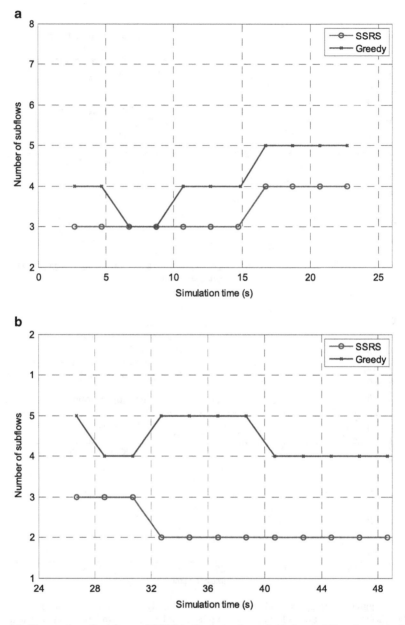

Fig. 4.6 The number of subflows of SSRS in the dynamic available bandwidth patterns at relays.
(**a**) Decreasing available bandwidth at relays. (**b**) Increasing available bandwidth at relays

Similar to the static scenario, Fig. 4.5 also shows that the goodput of SSRS and
Greedy are both much lower than their aggregate throughput. When the available
bandwidth is decreasing, both algorithms need to engage more relays to ensure

an aggregate throughput within the TBR range. This leads to a larger variation of the end-to-end path delays of the subflows. As a result, the goodput of both algorithms is degraded. Nonetheless, as seen in Fig. 4.6, SSRS uses less subflows in both decreasing and increasing scenarios, since it considers the number of relays in the priority index for relay ranking. Thus, SSRS achieves a goodput which is 50 % higher than that of Greedy.

4.3 Summary

In a user cooperation scenario, it is challenging to maintain a stable aggregate throughput at the destination, due to various factors, such as channel fading and local traffic fluctuation at relays. This chapter introduces a subset-sum based relay selection (SSRS) module based on a fully polynomial-time relay selection algorithm. By monitoring the available bandwidth variations, SSRS effectively adapts the relay paths in a highly dynamic environment so as to satisfy the application throughput requirement defined in terms of TBR. Specifically, the aggregate throughput at the destination is maintained within an acceptable TBR range via adding and deleting relay paths. While the best relay set is updated periodically, an active set is migrated to the backup set whenever the aggregate throughput is observed out of the TBR range.

Based on extensive simulations via ns-3 [5] for varying background traffic patterns, it is clearly shown that SSRS achieves a stable aggregate throughput with less performance outage and variation by engaging a much smaller number of subflows, compared to the greedy relay selection algorithm. The simulation results also show that the goodput of MPTCP with SSRS in the cooperation scenario is much lower than the aggregate throughput due to disparate end-to-end path delays (e.g., the goodput is only 55 % of aggregate throughput on average), which can cause the out-of-order issue and jeopardize the goodput at the destination. Chapters 5 and 6 further investigate how to improve the goodput.

References

1. Cormen, T.H., Leiserson, C.E., Rivest, R.L., Stein, C.: Introduction to Algortihms. The MIT Press (2009)
2. Ford, A., Raiciu, C., Handley, M., Barre, S., Iyengar, J.: Architectural guidelines for multipath TCP development. IETF RFC 6182 (2011)
3. Hulu: Hulu.com system requirements. http://www.hulu.com/help (2013)
4. Netflix: Internet connection speed recommendations. https://help.netflix.com/help (2013)
5. NS-3: The network simulator - ns-3. http://www.nsnam.org/ (2013)
6. Scharf, M., Ford, A.: MPTCP application interface considerations. IETF draft-ietf-mptcp-api-07 (2013)

7. Werle, J.: Bandwidth requirements by application. https://wiki.internet2.edu/confluence/display/k20t/Bandwidth+Requirements+by+Application (2013)
8. YouTube: System requirements. https://support.google.com/youtube/answer/78358?hl=en (2013)
9. Zhou, D., Song, W., Ju, P.: Subset-sum based relay selection for multipath TCP in cooperative LTE networks. In: Proc. IEEE GLOBECOM (2013)
10. Zhou, D., Song, W., Ju, P.: Goodput improvement for multipath transport control protocol in cooperative relay-based wireless networks. IET Communications **8**(9), 1541–1550 (2014)

Chapter 5
Adaptive Congestion Control (ACC)

Although SSRS [3] can guarantee a stable aggregate throughput with MPTCP in the user cooperation scenario, the goodput of MPTCP is still far lower than the aggregate throughput because the end-to-end delay differences of paths can cause out-of-order packets. This chapter introduces an adaptive congestion control (ACC) algorithm at the source [5], which dynamically adjusts the congestion window for each subflow TCP so as to mitigate the variation of the end-to-end path delay. The simulation results show that ACC together with SSRS can greatly improve the goodput performance of MPTCP in both the static and dynamic scenarios.

5.1 Goodput Analysis[1]

Although SSRS is designed to enable a stable aggregate throughput, it is the goodput that reflects the real application-level throughput. Here, the goodput is the amount of useful data available to the receiver application per unit time. In other words, goodput is the effective throughput perceived by the application. For example, if a user downloads a file from a file transfer protocol (FTP) server, the goodput that the user experiences is the file size divided by the file transfer time. The end-to-end delays of different paths can be considerably different, so the data sequence numbers (DSN) [1] of the received packets of each subflow may be out-of-order at the destination, which further harms the goodput.

As shown in Fig. 5.1, the source transmits video data to the destination by MPTCP with two subflows, which run over path 1 and path 2, respectively. The end-to-end delay of path 1 is half of that of path 2. Considering the case that the source

---

[1] ©IEEE. Reprinted with permission, from Proceedings of IEEE CCNC 2013, "Goodput improvement for multipath TCP by congestion window adaptation in multi-radio devices," by D. Zhou, W. Song, and M. Shi [5].

© The Author(s) 2014
D. Zhou, W. Song, *Multipath TCP for User Cooperation in Wireless Networks*,
SpringerBriefs in Computer Science, DOI 10.1007/978-3-319-11701-0_5

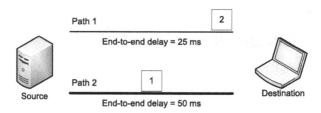

Fig. 5.1 An example of goodput

sends packets 1 and 2 simultaneously over path 2 and path 1, respectively, packet 2 will arrive at the destination first. If these two packets belong to the same video frame, then the destination cannot play out the frame until it receives packet 1. As the destination needs more time to receive the entire frame, the goodput is degraded.

In the following, two special scenarios of MPTCP are first examined so as to find out the primary factors affecting the goodput performance of MPTCP. Based on the analysis results, an adaptive congestion control (ACC) algorithm at the source is introduced in Sect. 5.2 to enhance the goodput of the destination. In this brief, the goodput of MPTCP is defined as the effective throughput of in-order packets delivered by MPTCP to the application layer, that is,

$$Goodput = \frac{\text{Payload size of } M \text{ in-order packets}}{\text{Total receiving time of } M \text{ packets}}. \tag{5.1}$$

Suppose that there are two available paths via two network interfaces of the source and destination interface 1 (IF1) and interface 2 (IF2). Let τ_i denote the packet sending interval at the source for path i, $i = 1, 2$. Assume that the throughput of path 1 is greater than that of path 2. Denote the end-to-end delay of path i by d_i, and assume $d_1 < d_2$. Consider a block of M packets with continuous DSN numbers, among which $M - 1$ packets are received on path 1 and only 1 packet is from path 2. Such a block of data packets is referred to as an *in-order unit*. Let S and T denote the total size in the unit of maximum segment size (MSS), which is the largest amount of data that a computer or communications device can receive in a single TCP segment, and the total receiving time of an in-order unit, respectively. Then, the goodput is evaluated by $G = S/T$.

Consider two special cases illustrated in Fig. 5.2. The in-order unit comprises 4 packets of DSN numbers 1, 2, 3, and 4. Suppose that packet 1 and packet 2 are sent at the same time to path 1 and path 2, respectively. Figure 5.2a shows the case with $\Delta D \triangleq d_2 - d_1 > \tau_1$. As shown in Fig. 5.2a, $T = \Delta D$ in this case. The goodput is thus expressed as

$$G = \frac{S}{T} = \frac{\frac{\tau_2}{\tau_1} + 1}{\Delta D}. \tag{5.2}$$

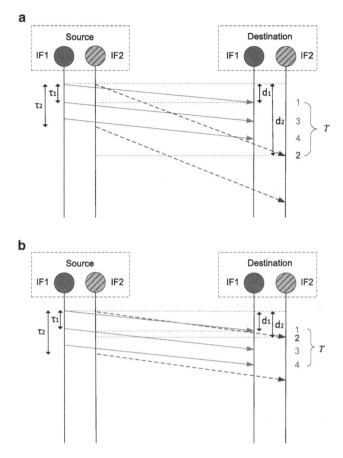

Fig. 5.2 Special cases with two transmission paths for goodput analysis. (**a**) General case with $\Delta D > \tau_1$. (**b**) Near optimal case with $\Delta D \leq \tau_1$

Equation (5.2) implies that goodput is inversely proportional to the end-to-end path delay difference ΔD. The larger difference of the end-to-end path delays, the more out-of-order packets that will be received at the destination, which in turn decreases the goodput. Therefore, it is important to minimize the end-to-end delay difference among paths.

Figure 5.2b shows another special scenario with $\Delta D \leq \tau_1$. In this case, the destination needs less time to receive all packets within the in-order unit. Here, the total time to receive all M packets of the in-order unit is just the time for path 1 to receive all $M - 1$ packets sent over it. As shown in Fig. 5.2b, $T = \tau_2$ in this case. The goodput is then obtained as

$$G = \frac{S}{T} = \frac{\frac{\tau_2}{\tau_1} + 1}{\tau_2} = \frac{1}{\tau_1} + \frac{1}{\tau_2}. \tag{5.3}$$

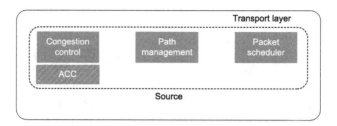

Fig. 5.3 Structure of ACC

The aggregate throughput over two paths (in the unit of MSS per second) in an in-order unit is given by

$$\Upsilon = \frac{1}{\tau_1} + \frac{1}{\tau_2}. \tag{5.4}$$

Obviously, $\Upsilon = G$ in the special case of Fig. 5.2b. This means that, when the delay difference of the paths is small and the packets are properly scheduled over the paths, the goodput can approach its upper bound, i.e., the aggregate throughput.

5.2   ACC Module[2]

In regular TCP, the source maintains a congestion window, which limits the maximum number of packets that can be sent on the fly without received ACKs. Once triple duplicate ACKs are received, the source interprets this event as an indicator of packet loss and halves its congestion window to reduce the traffic load toward the corresponding transmission path [2]. As discussed in Chap. 2, each subflow TCP of MPTCP maintains its own congestion window. These congestion windows are increased cooperatively by a coupled congestion control algorithm. In contrast, the congestion windows are decreased independently. Specifically, when the subflow TCP at the source receives the congestion signals, e.g., three duplicated ACKs, it decreases the congestion window of this path by half. Consequently, the congestion window of each path may greatly differ from each other because different paths have different congestion situations. This will lead to a large path delay difference, which is detrimental to the goodput performance.

This section introduces an ACC algorithm to improve the goodput performance of MPTCP. As shown in Fig. 5.3, ACC complements the congestion control algorithm of MPTCP. In ACC, the average end-to-end delay of each path is monitored

---
[2]Reprinted with permission, from Proceedings of IEEE CCNC 2013, "Goodput improvement for multipath TCP by congestion window adaptation in multi-radio devices," by D. Zhou, W. Song, and M. Shi [5].

Algorithm 3 Adaptive congestion control

1: **Input**: End-to-end delay of paths d_r, $\forall 1 \le r \le N_s$; range of delay ratio $[\eta_{min}, \eta_{max}]$; current
    system time T_{cur}; creation time of subflow over of slowest path l, T_{init}^{l}; adaptation start time
    Γ; maximum adaptation limit m
2: **Output**: Congestion window of slowest path j, $cwnd_j$; slow start threshold of slowest path
    j, $ssthresh_j$
3: $j = \arg \max\limits_{1 \le r \le N_s} (d_r)$ // Find the slowest path j
4: $l = \arg \min\limits_{1 \le r \le N_s} (d_r)$ // Find the fastest path l
5: $\eta = d_j / d_l$
       // High delay ratio detected, select the slowest path j for $cwnd$ adaptation
6: **if** $\eta_{min} \le \eta \le \eta_{max}$ **then**
       // Adaptation counter does not exceed the maximum limit
7:     **if** $count_j < m$ **then**
8:       $elapse_j = T_{cur} - T_{init}^{l}$
9:       **if** $elapse_j > \Gamma$ **then**
            // Decrease the congestion window of path l
10:         $cwnd_j \leftarrow cwnd_j / \eta$
11:         **if** $ssthresh_j > cwnd_j$ **then**
12:           $ssthresh_j = cwnd_j$
13:         **end if**
14:         $count_j \leftarrow count_j + 1$
15:       **end if**
16:     **else**
          // Reset adaptation counter
17:       $count_j = 0$
18:     **end if**
19: **end if**

---

at certain frequency, which is also the frequency of the periodic relay bandwidth broadcast for SSRS. Among all the used paths, the ratio of the maximum path delay over the minimum path delay is defined as *delay ratio*. When a large delay ratio is detected, the source proportionally decreases its congestion window even when there is no congestion signal detected. The main purpose is to minimize the path delay difference to increase the goodput. On the other hand, the increase of the subflow congestion window follows the original congestion control algorithm of MPTCP.

The details of ACC are given in Algorithm 3. As seen, the ACC algorithm is triggered when the delay ratio η is within a certain range $[\eta_{min}, \eta_{max}]$. Here, the delay ratio η is defined as $\eta = \max_{1 \le r \le N_s} d_r / \min_{1 \le r \le N_s} d_r$, where N_s is the number of paths, and d_r is the end-to-end delay of path r. In Line 6, the congestion window (denoted by $cwnd_r$) of path r with the maximum delay is decreased proportionally by the delay ratio η. This is because a larger delay ratio indicates that the high delay path is overloaded. Its congestion window needs to be decreased to relieve the traffic load and reduce the end-to-end path delay.

Here, η_{max} is introduced to avoid over blocking the slow path. Since decreasing the congestion window of a path too much will halt the source from sending packets to the subflow on that path in a long time, which will lower the throughput on that path and thus jeopardize the aggregate throughput.

Meanwhile, since ACC only works in congestion avoidance stage, it tries to avoid the slow start during adjusting congestion window. In ACC, the TCP slow start threshold of the slowest path j (denoted by $ssthresh_j$) need to be set to the new $cwnd_j$, if $ssthresh_j > cwnd_j$. If $ssthreh_j$ is not updated in this case, $cwnd_j$ will be recovered quickly with the slow start procedure, i.e., $cwnd_j$ is linearly increased by 1 for each successful ACK received at the source. Due to the unwanted slow start procedure, the congestion window of the slow path is not decreased for sufficient time to reduce the end-to-end path delay.

Although adjusting $cwnd$ can reduce the end-to-end delay variation of multiple paths, it is hard to guarantee that all paths maintain similar delays as the ideal case illustrated in Fig. 5.2b. This is due to a variety of factors in addition to the traffic load over the paths that contribute to the end-to-end path delay, such as transmission, processing, and queueing delays at routers, base stations, and intermediate nodes between communication peers. The link-layer interference and retransmission over wireless channels also result in a path delay variation. The transport-layer control itself cannot completely eliminate the path delay variation. A parameter $count_j$ is introduced to restrict the number of continuous reductions of congestion window for a single path j by m, which is the maximum adaptation limit, e.g., $m = 3$ in the experiments. As such, severe throughput degradation on an individual path can be avoided. In addition, ACC is not activated before the subflow TCP has run for a period Γ. Threshold Γ is needed because the end-to-end delay measurements for a new path cannot accurately reflect the real congestion situation immediately. In practice, the time threshold Γ can be set to several times of the maximum RTT to make sure that the subflow TCP to adjust is already in the congestion avoidance stage.

5.3   Experimental Results of ACC with and w/o SSRS[3]

The simulation results in Chap. 4 show that SSRS can effectively guarantee a stable aggregate throughput to satisfy the TBR requirement. Chapter 4 results also show that the goodput, which reflects the effective throughput perceived at the application layer, is much lower than the aggregate throughput. This section evaluates the goodput of MPTCP with ACC upon SSRS in various patterns of background traffic load at relays. The traffic load patterns considered here are the same as those used in Chap. 4. The maximum and minimum delay ratios (i.e., η_{max} and η_{min}) are set to 3 and 1, respectively. The source measures the end-to-end path delay via N relays every 2 s, which is the same as the time period that SSRS broadcasts the available

---

[3]Reprinted with permission from IET Communications (2014), "Goodput improvement for multipath transport control protocol in cooperative relay-based wireless networks," by D. Zhou, W. Song, and P. Ju [4].

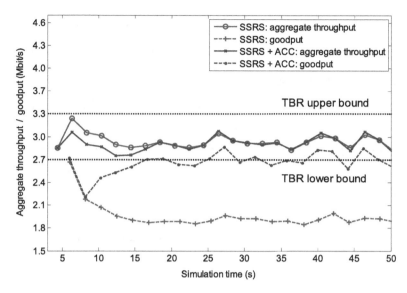

Fig. 5.4 Aggregate throughput and goodput of ACC in the static available bandwidth pattern at relays

bandwidth of each relay. The topology and distribution of the relays also follow the configuration considered in Sect. 4.2. The other default simulation parameters are given in Table 3.1.

Figure 5.4 shows the aggregate throughput and the goodput of ACC with SSRS in the static available bandwidth pattern. It is clearly seen that ACC can effectively enhance the goodput of MPTCP at the destination, as compared to SSRS alone. Specifically, the average goodput of ACC together with SSRS is 1.42 times higher than that of SSRS alone. This is because ACC dynamically adapts the congestion window ($cwnd$) of each path so as to mitigate the delay differences of relay paths.

Another interesting observation of Fig. 5.4 is that, between the 4th s and the 16th s, the aggregate throughput of ACC with SSRS is slightly smaller than that of SSRS alone. Actually, at the 2nd s, three new relays are bound to the destination and lead to an increasing delay difference from that moment. At the 4th s, a larger delay ratio is detected within the range $[\eta_{min}, \eta_{max}]$. ACC is then triggered to decrease the $cwnd$ of the slow path and relieve its traffic load. As a result, there is a slight throughput degradation for a brief period (12 s) following that between the 4th s and the 16th s. In a nutshell, ACC may temporarily slow down or even suspend the source for a short period by decreasing the congestion window to reduce the path delay. As a side effect, the aggregate throughput may be slightly degraded by ACC. To avoid penalizing a slow path frequently, a maximum adaptation limit m is used in ACC to restrict the number of times the $cwnd$ of one subflow can be decreased. Although ACC slightly decreases the aggregate throughput in certain cases, the overall goodput is still much larger than that of SSRS alone.

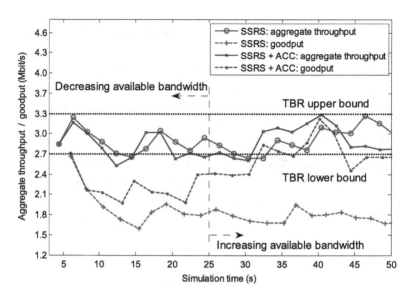

Fig. 5.5 Aggregate throughput and goodput of ACC in the dynamic available bandwidth patterns at relays

Figure 5.5 shows the aggregate throughput and the goodput of ACC with SSRS in the dynamic available bandwidth patterns. It can be seen that ACC achieves a higher goodput in both the decreasing and increasing available bandwidth patterns as compared to SSRS. Specifically, on average, the goodput of SSRS + ACC is 1.33 higher larger than that of SSRS alone. Moreover, the goodput of ACC in the decreasing available bandwidth pattern is smaller than that in the increasing available bandwidth pattern.

There are two main reasons behind the observation. First, when the available bandwidth is decreasing, the end-to-end delays of some paths become so large that the delay ratio exceeds η_{max} (set to 3 in the simulation). For example, the maximum delay ratio between the 5th s and the 25th s is observed to be 5, which is greater than the maximum delay ratio 3. Also, ACC itself cannot significantly reduce such a large delay difference among the paths, since many other factors in addition to traffic load (controlled by the congestion window of a path) can contribute to the delay, e.g., queueing delay at routers.

Second, ACC slightly degrades the aggregate throughput at the destination by shrinking the congestion window. In the decreasing available bandwidth pattern, the aggregate bandwidth of SSRS is already very close to the lower bound of TBR, e.g., at the 15th s and the 23rd s. When ACC is used with SSRS, more outage events occur. Between the 5th s and 25th s, there is 1 outage event in SSRS, while there are 4 throughput outages with ACC. Such outages result in frequent relay updates in SSRS. Some subflows need to be deleted or added in a high frequency so that the end-to-end delay of the path is decreased. As a result, the effectiveness of ACC is further constrained. For instance, the shortest connection time between a relay and the destination in the simulation is 2 s, which is only three times that of the RTT of the path. This connection time is too short for ACC adaptation to take effect and reduce the end-to-end path delay.

5.4 Summary

This chapter introduces an adaptive congestion control (ACC) module to enhance the goodput performance of MPTCP at the destination. By adapting the congestion window based on the end-to-end delay differences among paths, ACC effectively improves the goodput of MPTCP by running with SSRS. Simulation results demonstrate that the goodput of MPTCP is significantly improved by using ACC together with SSRS. On average, the goodput of SSRS + ACC is 1.5 times and 1.33 times larger than that of SSRS alone in static pattern and dynamic pattern, respectively. Many factors contribute to the end-to-end path delay, such as the queue length at the routers and packet retransmission over wireless links, so ACC cannot completely eliminate the delay gap of the paths. As a result, the goodput is not so stable in a highly dynamic environment with a large variation for the background traffic load as that in the static scenario. The next chapter introduces another module which can work cooperatively with ACC to achieve a more stable goodput for MPTCP in a user cooperation scenario.

References

1. Ford, A., Raiciu, C., Handley, M., Bonaventure, O.: TCP extensions for multipath operation with multiple addresses. IETF RFC 6824 (2013)
2. Mathis, M., Mahdavi, J., Floyd, S., Romanow, A.: TCP selective acknowledgment options. IETF RFC 2018 (1996)
3. Zhou, D., Song, W., Ju, P.: Subset-sum based relay selection for multipath TCP in cooperative LTE networks. In: Proc. IEEE GLOBECOM (2013)
4. Zhou, D., Song, W., Ju, P.: Goodput improvement for multipath transport control protocol in cooperative relay-based wireless networks. IET Communications 8(9), 1541–1550 (2014)
5. Zhou, D., Song, W., Shi, M.: Goodput improvement for multipath TCP by congestion window adaptation in multi-radio devices. In: Proc. IEEE CCNC (2013)

Chapter 6
Differentiated Packet Forwarding (DPF)

As discussed in Chap. 5, because many factors contribute to the end-to-end path delay, ACC cannot eliminate the delay difference among paths. This chapter introduces an extension module, namely, differentiated packet forwarding (DPF) [5], for MPTCP [3] in a user cooperation scenario to further improve the goodput at the destination. DPF takes advantage of the relays to buffer out-of-order packets and only forwards the packets in an expected DSN range to the destination. A fast ACK scheme can be used at the relays to return ACK messages to the source on behalf of the destination so as to reduce the ACK delay and avoid the unnecessary decreasing of the congestion window caused by packet buffering at the relays. The simulation results show that DPF significantly improves the goodput of MPTCP. In addition, the overall performance with ACC [8], DPF [5] and SSRS [6] working together is evaluated. It is found that the three modules are complementary and can achieve the best goodput performance.

6.1 DPF Module[1]

Figure 6.1 shows the structure of differentiated packet forwarding (DPF) in the rectangles at the transport layer. In Fig. 6.1, DPF manager at the destination informs the DPF client at each relay an expected range of MPTCP data sequence numbers (DSN) Then, the relays buffer the packets outside the range until they are requested by the destination via a range update message. Only the in-range packets are forwarded to the destination. By this means, the goodput can be improved and the receiving buffer of the relays can also be used by the destination.

---

[1] ©IEEE. Reprinted with permission, from Proceedings of IEEE PIMRC 2012, "Performance enhancement of multipath TCP with cooperative relays in a collaborative community," by D. Zhou, P. Ju, and W. Song [5].

© The Author(s) 2014
D. Zhou, W. Song, *Multipath TCP for User Cooperation in Wireless Networks*,
SpringerBriefs in Computer Science, DOI 10.1007/978-3-319-11701-0__6

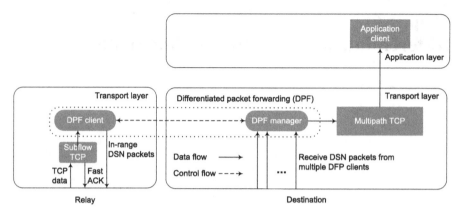

Fig. 6.1   Structure of DPF

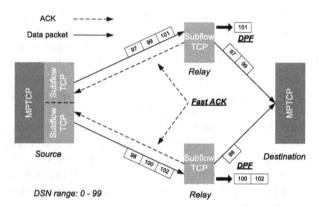

Fig. 6.2   An illustration of fast ACK and DPF

One side effect of buffering packets at the relays is that the source may unnecessarily decrease the congestion window, since the ACKs of the buffered packets are delayed, which may cause out-of-order ACKs that will be interpreted as an indicator of packet loss by the source. To address this problem, a fast ACK scheme can be further used to migrate the endpoints of the subflow TCP from the destination to the relay, so that the relay can return ACKs on behalf of the destination. Figure 6.2 shows an example of fast ACK and DPF. Given an expected DSN range [0,99], the relays only forward the packets of a DSN within this range to the destination. Meanwhile, the relays buffer all the packets whose DSN is out of this range, e.g., packets 100, 101, and 102. All ACKs are returned to the source by the relays instead of the destination. To guarantee that the packets successfully arrive at the destination eventually, the relays require positive acknowledgements from the destination for the forwarded packets via regular TCP. The relays will buffer these packets and retransmit lost packets until positive acknowledgements are returned from the destination.

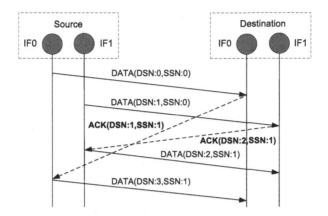

Fig. 6.3 MPTCP sequence numbers DSN and SSN for acknowledgement

In the rest of this section, the procedures of the fast ACK scheme are first introduced. Then, the DPF algorithms at the relays and the destination are illustrated.

6.1.1 Fast ACK

As introduced in Chap. 2, MPTCP has two levels of sequence number space: SSN and DSN. Correspondingly, MPTCP has two types of ACK messages, *Subflow ACK* and *Data ACK* [2]. In the subflow TCP, the acknowledgement number in Subflow ACK specifies the next in-order SSN that the destination expects to receive. In contrast, the acknowledgement DSN number in Data ACK only indicates that the destination has successfully received the packets whose DSNs are less than the DSN in Data ACK. The source sends the next packet based on a scheduler rather than the DSN number in Data ACK. Otherwise, duplicate transmissions are possible to cause waste of bandwidth resources.

Figure 6.3 shows how the sequence numbers are used in ACK messages. Here, the data are sent between the source and the destination over two paths via two network interfaces IF0 and IF1. Data ACK for packet 0 arrives at the source after Data ACK for packet 1. Once receiving the Data ACK for packet 0 at the interface 0 (IF0), the source immediately sends out packet 3 instead of packet 1. As introduced in Sect. 2.3.1, the SSN is similar to the sequence number used in regular TCP. An ACK message of an SSN h over a path indicates all packets sent over this path with an SSN less than h have been successfully received. The next expected packet over this path should take an SSN of h. In contrast, the DSN l in an ACK message only indicates the packet of a DSN $l - 1$ has been received. It is up to the scheduler at the source to determine which packet should be sent next over a path. The packet is not necessary the one with a DSN l as in regular TCP. In Fig. 6.3, once the ACK with a DSN 1 is received at the source, the source sends out the packet of a DSN

3 instead of a DSN 1 over that path with IF0. This is because the source knows the packet of DSN 1 has been received over the path via IF1 by using the unique DSNs of packets. Duplicate transmissions can thus be avoided so that it is possible to allow multiple relays to return Data ACK messages independently on behalf of the destination.

In order to provide timely ACK feedback to the source, MPTCP can be extended to migrate the connection endpoint of subflow TCP from the destination to relays. After setting up the first subflow TCP with the source, the destination sends an address advertisement to the source with an `Add Address` option [1], which contains the address and the port number of one or multiple relays. Then, the source initiates a "SYN-SYN/ACK-ACK" three-way handshaking and establishes one or multiple subflow TCP connections with the relays. Once receiving packets from the relays, the destination knows that the new path has been established between the source and the relay. It then sends a message to the source, which contains a `Remove Address` option to delete the original subflow between the source and the destination. As such, the relays are visible to the source as the endpoints of the end-to-end connection. The original endpoint at the destination is deleted so that the relays as the new endpoints take the responsibility of returning ACKs on behalf of the destination. As seen, the migration of the endpoints form the destination to the relays reuses the standard MPTCP signalling options such as `Add Address` and `Remove Address`. This makes easy deployment of the Fast ACK mechanism in practice.

The above procedure exploits the signalling of MPTCP to add relays and enable fast ACK via relays. Nonetheless, the security mechanism of MPTCP [4] poses another technical issue that needs to be handled properly. According to MPTCP [2], the connection initiation begins with a `SYN`, `SYN/ACK`, `ACK` exchange on a single path. This MPTCP handshake negotiates the cryptographic algorithm and declares the sender's and receiver's 64-bit keys, which are used to authenticate the addition of future subflows. The cryptographic hash of the key is a 32-bit *token* as a local unique connection identifier. When a new subflow is started, the SYN, SYN/ACK and ACK packets also carry an `MP_JOIN` option, which contains the receiver's token to identify the MPTCP connection it is joining, and the sender's 64-bit truncated message authentication code (MAC) for authentication. As seen, the relays joining the MPTCP connection need to know the tokens of the source and the destination, which are generated during the first subflow establishment. Also, the destination should obtain the addresses and port numbers of the relays before sending an address advertisement to the source with an `Add Address` option. Hence, it is required that the destination and relays exchange address information and tokens in advance via additional signalling.

Algorithm 4 Differentiated packet forwarding at the destination

---
Input: ρ, MIN_DSN, MAX_DSN, H, H_{rcv}
Output: H_{rcv}, MIN_DSN, MAX_DSN
 1: SendDsnRange(MIN_DSN, MAX_DSN)   // Send the DSN range to relays
 2: **while** a new packet of DSN is received **do**
 3:     **if** $MIN\_DSN \leq DSN \leq MAX\_DSN$ **then**
 4:         $H_{rcv} \leftarrow H_{rcv} + 1$
 5:     **end if**
        // Accept and append new packet to receive buffer
 6:     AddToBuffer()
 7:     **if** $H_{rcv}/H \geq \rho$ **then**
 8:         $MIN\_DSN \leftarrow MIN\_DSN + H$
 9:         $MAX\_DSN \leftarrow MAX\_DSN + H$
        // Send update messages for DSN range to the relays
10:         SendDsnRange(MIN_DSN, MAX_DSN)
11:     **end if**
12: **end while**

6.1.2   Differentiated Packet Forwarding

Algorithms 4 and 5 give the details of DPF, which runs in a distributed fashion at the destination and the relays. Algorithm 4 shows DPF on the destination side. Once all MPTCP connections between the source and the relays are established, the DPF manager at the destination sends the expected DSN range, (MIN_DSN, MAX_DSN), for the incoming data packets to the relays.

The length of the DSN range is H packets. The value of H depends on many factors such as the out-of-order level and delays of different paths. When DPF is working together with SSRS, a stable aggregate throughput around TBR can be achieved by selecting and switching relay sets. Further considering the available bandwidth monitoring period (in seconds) F of SSRS, a reasonable value $H = TBR \cdot F$ is chosen. If any packet within the DSN range is received, the destination accepts and buffers the packets. Meanwhile, the destination counts the number of received packets of continuous DSN falling in the expected range, denoted by H_{rcv}. The ratio H_{rcv}/H indicates the occupancy level of the receive buffer and the completeness of an in-order unit. Given a threshold ρ ($0 < \rho \leq 1$), when $H_{rcv}/H \geq \rho$, the destination sends update messages for the new DSN range to the relays. The value of ρ can be less than 1 so that a margin time is reserved for the relays to forward buffered packets to the destination. The effect of ρ is further discussed in Sect. 6.2.2.

Algorithm 5 defines the DPF algorithm working at each relay. Once a packet within the expected DSN range is received at a relay, it forwards the packet toward the destination. If the packet DSN is smaller than the lower bound of DSN range MIN_DSN, it indicates an urgent out-of-order packet belonging to previous DSN ranges. The relay also forwards such a packet immediately. In contrast, if the DSN

Algorithm 5 Differentiated packet forwarding at the relay

Input: ρ, MIN_DSN, MAX_DSN, MAX_BF_SIZE
Output: null
   // Receive the first DSN range from the destination
 1: ReceiveDsnRange(MIN_DSN, MAX_DSN)
 2: **for** each packet of DSN in receive buffer **do**
 3:    **if** $MIN\_DSN \leq DSN \leq MAX\_DSN$ **then**
        // Forward the buffered packet toward the destination
 4:       Forward(buffered packet)
 5:    **end if**
 6: **end for**
 7: **while** a new packet of DSN is received **do**
 8:    **if** $MIN\_DSN \leq DSN \leq MAX\_DSN$ or
 9:    $DSN < MIN\_DSN$ **then**
        // Forward new packet toward the destination
10:      Forward(new packet)
11:    **else**
12:      **if** $DSN > MAX\_DSN$ **then**
13:        **if** Buffer length $\geq MAX\_BF\_SIZE$ **then**
           // Forward head packet of receive buffer to the destination
14:         Forward(head packet)
15:        **end if**
        // Append new packet to the end of receive buffer
16:        AddToBuffer()
17:      **end if**
18:    **end if**
19: **end while**

---

of a received packet is larger than the upper bound MAX_DSN and the receive buffer is not full, the relay buffers and retains the packet on behalf of the destination. Meanwhile, if the receive buffer overflows, the relay forwards the packet in the head of queue to the destination. The relay also needs to buffer the packets that have been forwarded to the destination but without positive acknowledgements from the destination yet. Since the relay also has its own local traffic, it is necessary to limit the maximum buffer size it can provide to the destination, denoted by MAX_BF_SIZE. Obviously, MAX_BF_SIZE should be greater than H.

Once an update message for DSN range is received, the relay forwards all the buffered packets within the new DSN range to the destination. The DSN range update is based on the threshold ρ maintained at the destination, which indicates the occupancy level of receive buffer and the completeness of a consecutive data block (an in-order unit).

6.2 Experimental Results of DPF Combined with SSRS and/or ACC[2]

Simulation results in Chap. 5 show that ACC working together with SSRS dramatically improves the goodput performance at the destination. The goodput of ACC with SSRS is not so stable in a highly dynamic environment, e.g., with a large variation for the background traffic load. This is because ACC cannot eliminate the delay gap of the paths due to many factors contributing to the end-to-end delay, such as the queue length at the routers and packet retransmission over wireless links. This section evaluates the performance of DPF with SSRS in various scenarios. The update threshold ρ is set to 0.95 by default. The overall goodput performance with ACC, DPF and SSRS working together is also examined. The spatial distribution and background traffic load of the relays are the same as the setup considered in Sect. 4.2. The other simulation parameters follow the default parameters given in Table 3.1.

6.2.1 Differentiated Packet Forwarding

Figure 6.4 shows the aggregate throughput and the goodput of DPF with SSRS in the static available bandwidth pattern. It is clearly seen that DPF dramatically improves the goodput, as compared to SSRS. The average goodput of DPF together with SSRS is 1.36 times larger than that of SSRS alone, since the number of out-of-order packets at the destination is decreased by buffering the out-of-order packets at the relays.

Figure 6.5 shows the aggregate throughput and the goodput of DPF with SSRS in the dynamic available bandwidth patterns. It can be seen that the goodput of DPF is 1.6 times larger than that of SSRS alone on average than that of SSRS in both the decreasing and increasing available bandwidth patterns. When the available bandwidth is decreasing, the end-to-end delays of some paths become so large that an increasing number of out-of-order packets are buffered at the relays, which helps improve the goodput performance at the destination. Meanwhile, more buffer space is needed at the relays to accommodate the out-of-order packets.

Figure 6.5 also shows that the aggregate throughput of DPF with SSRS is more stable than that of ACC with SSRS in Fig. 5.5. The relays return Data ACK and Subflow ACK to the source on behalf of the destination, so DPF is transparent to the source so that the sending rate at the source will not be affected by DPF.

---

[2]Reprinted with permission from IET Communications (2014), "Goodput improvement for multipath transport control protocol in cooperative relay-based wireless networks," by D. Zhou, W. Song, and P. Ju [7].

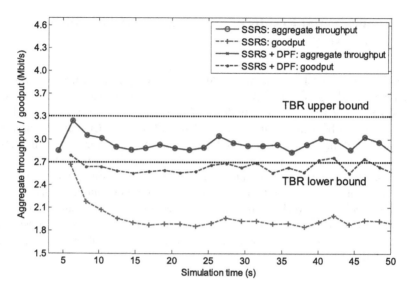

Fig. 6.4 Aggregate throughput and goodput of DPF in the static available bandwidth pattern at relays

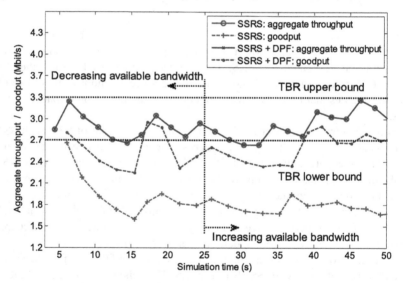

Fig. 6.5 Aggregate throughput and goodput of DPF in the dynamic available bandwidth patterns at relays

6.2.2 Overall Performance Evaluation

The simulation results in Sects. 6.2.1 and 5.3 show that ACC and DPF working together with SSRS can achieve a higher goodput than SSRS alone in both the static and dynamic available bandwidth patterns. Nevertheless, when the available bandwidth is decreasing, ACC cannot guarantee a stable goodput, while DPF costs more buffer space at relays. This section evaluates the overall performance with the three modules. It will be seen in the following results that ACC and DPF can complement the limitations of each other in the low available bandwidth scenario.

To combine the three modules, SSRS is first activated at the relays and the destination to achieve a stable aggregate throughput to satisfy the TBR requirement. Also, ACC runs at the source to enhance the goodput of SSRS. Moreover, DPF is triggered at the relays and the destination to cooperatively work with ACC to further improve the goodput at the destination.

Figure 6.6 shows the goodput of the three combined modules (SSRS + ACC + DPF) in both the static and dynamic available bandwidth patterns. In the experiment, the TBR and variation ratio θ is set to 3.0 Mbit/s and 0.3 respectively, which are the same as in Chap. 4. In Fig. 6.6a, it is clearly seen that the goodput of the three combined modules performs the best among the other combinations in the static pattern. Particularly, the overall goodput is much more stable than that of ACC with SSRS at the beginning of the MPTCP connection, e.g., from the 5th s to the 10th s. This is because the out-of-order packets caused by mismatched end-to-end path delays are buffered by the DPF module at the relays. In contrast, ACC is not activated during this time period, since the end-to-end delay measurements for a new path are not completed yet and cannot accurately reflect the real congestion situation. Hence, the goodput of ACC with SSRS is almost the same as that of SSRS alone at the beginning of the MPTCP connection.

In the dynamic patterns, as shown in Fig. 6.6b, the goodput of the three combined modules outperforms ACC with SSRS in all monitor time points, and exceeds the goodput of DPF with SSRS in the decreasing and increasing available bandwidth patterns in 50 % and 91 % of the monitor time points, respectively. In some cases of the decreasing available bandwidth pattern, the three combined modules have a slightly lower goodput than that of DPF with SSRS. This is mainly because ACC degrades the aggregate throughput, which is the upper bound of goodput.

In the decreasing available bandwidth pattern, the end-to-end delay ratio of some paths is much larger than θ_{max} (set to 3). For example, a delay ratio as high as 5 is observed in the simulation. As a result, ACC takes effect and tries to reduce the large end-to-end path delay by shrinking the corresponding congestion window of the path, which degrades the aggregate throughput. Even though DPF improves the goodput more times in the decreasing available bandwidth pattern, the goodput of the three combined modules is still smaller than that of DPF with SSRS in some cases due to the smaller aggregate throughput.

Figure 6.7 shows the goodput of the three combined modules with a different DSN range update threshold ρ of DPF. It can be seen that, when the threshold ρ is decreased from 0.95 to 0.85, the goodput is degraded in both the static and dynamic patterns. For a smaller threshold, there will be faster updates for the DSN range and more overlapping among packets in two adjacent DSN ranges. As a result, more out-of-order packets are received at the destination, which degrades the goodput.

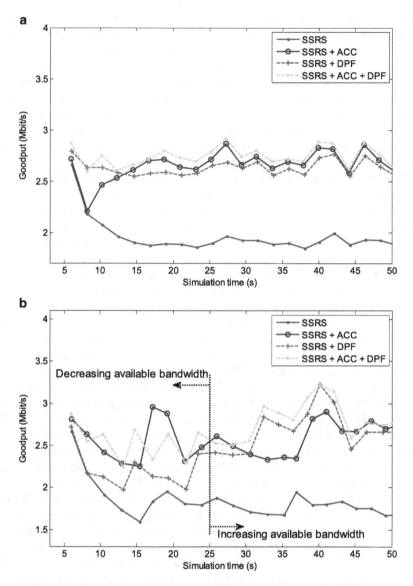

Fig. 6.6 Goodput of three combined modules (SSRS, ACC and DPF). (**a**) Static available bandwidth pattern at relays. (**b**) Dynamic available bandwidth pattern at relays

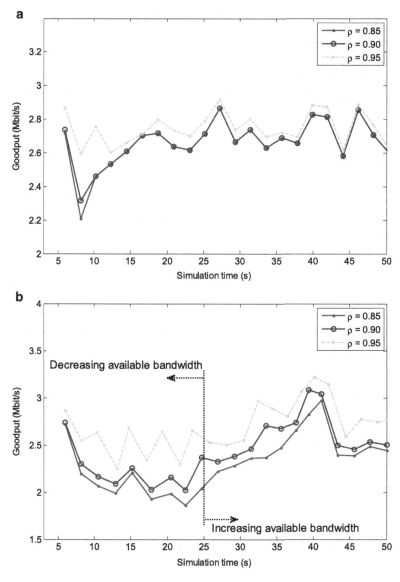

Fig. 6.7 Goodput of the three combined modules (SSRS, ACC and DPF) with different thresholds ρ. (**a**) Static available bandwidth pattern at relay. (**b**) Dynamic available bandwidth patterns at relay

Figure 6.7 also shows that the goodput decrease for ρ from 0.95 to 0.90 is smaller than that with ρ from 0.90 to 0.85. In the static available bandwidth pattern, when $\rho = 0.95$, the goodput is almost the same as that with $\rho = 0.90$. Based on the preceding analysis, it is known that a smaller ρ results in a larger overlapping time between two adjacent DSN ranges. Therefore, if ρ is small enough, such as 0.85 in Fig. 6.7a, the goodput of the three combined modules falls back to the same as that of ACC with SSRS.

6.3 Summary

This chapter introduces differentiated packet forwarding (DPF) module for MPTCP in a user cooperation scenario to further improve the goodput at the destination. DPF works in a distributed manner at the relays and the destination. In DPF, the destination periodically informs an expected DSN range to the relays. Then, the relay buffers out-of-order packets and only forwards the packets in an expected DSN range to the destination, so that the goodput is not degraded. Also, a fast ACK scheme can be used to enable the relays to return ACK messages to the source for the buffered out-of-order packets on behalf of the destination. As such, the source will not unnecessarily decrease the congestion window on the path. The simulation results demonstrate that the goodput of MPTCP is significantly improved and more stable with DPF. In addition, the overall performance of the three modules (SSRS, ACC and DPF) working together is evaluated. The simulation results show that they provide the best goodput performance in both the static and dynamic available bandwidth patterns when used together. The impact of the DSN range update threshold ρ is also investigated. It is observed that the goodput is slightly increased when the threshold becomes larger within a range close to 1.

References

1. Ford, A., Raiciu, C., Handley, M., Bonaventure, O.: TCP extensions for multipath operation with multiple addresses. IETF draft-ietf-mptcp-multiaddressed-08 (2012)
2. Ford, A., Raiciu, C., Handley, M., Bonaventure, O.: TCP extensions for multipath operation with multiple addresses. IETF RFC 6824 (2013)
3. Ford, A., Raiciu, C., Handley, M., Barre, S., Iyengar, J.: Architectural guidelines for multipath TCP development. IETF RFC 6182 (2011)
4. Scharf, M., Ford, A.: Multipath TCP (MPTCP) application interface considerations. IETF RFC 6897 (2013)
5. Zhou, D., Ju, P., Song, W.: Performance enhancement of multipath TCP with cooperative relays in a collaborative community. In: Proc. IEEE PIMRC (2012)
6. Zhou, D., Song, W., Ju, P.: Subset-sum based relay selection for multipath TCP in cooperative LTE networks. In: Proc. IEEE GLOBECOM (2013)
7. Zhou, D., Song, W., Ju, P.: Goodput improvement for multipath transport control protocol in cooperative relay-based wireless networks. IET Communications (2014). To appear
8. Zhou, D., Song, W., Shi, M.: Goodput improvement for multipath TCP by congestion window adaptation in multi-radio devices. In: Proc. IEEE CCNC (2013)

Chapter 7
Bandwidth Sharing for User Cooperation

As shown in Chaps. 4, 5, and 6, the three extension modules to MPTCP [2], SSRS [9], ACC [10] and DPF [7], can work in a complementary fashion to guarantee a stable aggregate throughput that satisfies the TBR requirement and achieves a high goodput. So far, the discussion focuses on UDP-based local traffic at the relays. In this chapter, experiments further show that, when the local traffic at the relays is TCP-based, the throughput of the local traffic at the relays can be severely degraded by forwarding the MPTCP traffic for the destination. This behavior can jeopardize the motivation of mobile users to engage in user cooperation. Hence, this chapter introduces a bandwidth sharing module, which aims to guarantee fair bandwidth sharing between the MPTCP subflows and the local single-path flows at the relays. In particular, this bandwidth sharing module extends the standard congestion control algorithm of MPTCP to the user cooperation scenario. Simulation results demonstrate that this module ensures that the throughput of the local traffic at the relays is not degraded by adding MPTCP subflows for the destination, and thus better promotes the relays to engage in user cooperation.

7.1 MCC Bandwidth Sharing with User Cooperation

The coupled congestion control algorithm [5] of MPTCP is introduced in Sect. 2.3.1. For easy comparison, the main points of this algorithm are highlighted in the following:

- Once the source receives an acknowledgement (ACK) from path r, it increases the congestion window of path r by $\min(a/w_{total}, 1/w_r)$; and
- Once the source receives a congestion signal from path r, it decreases the congestion window w_r of path r to $w_r/2$.

© The Author(s) 2014
D. Zhou, W. Song, *Multipath TCP for User Cooperation in Wireless Networks*,
SpringerBriefs in Computer Science, DOI 10.1007/978-3-319-11701-0_7

Here, w_r is the current congestion window size (in the unit of MSS) of subflow on path r, w_{total} is the total congestion window size of all subflows, given by $w_{total} = \sum_{r=1}^{K_s} w_r$, where K_s is the number of subflows, and a is an *aggressiveness* factor defined by

$$a = w_{total} \frac{\max_{1 \le r \le K_s} \frac{w_r}{RTT_r^2}}{\left(\sum_{r=1}^{K_s} \frac{w_r}{RTT_r}\right)^2} \tag{7.1}$$

In (7.1), RTT_r is the RTT of path r. The increment $\min(a/w_{total}, 1/w_r)$ for congestion window size aims to ensure that each MPTCP subflow does not increase its congestion window faster than a single-path TCP flow with the same window size.

This algorithm aims to ensure that a multipath flow should (i) perform at least as well as a single-path flow on the best path available to it, and (ii) take no more capacity than a single-path flow would obtain at maximum when experiencing the same loss rate. Basically, the objective (i) motivates users to run MPTCP, while the objective (ii) guarantees that an MPTCP flow gracefully shares the path bandwidth with regular single-path TCP flows.

The congestion control algorithm of MPTCP focuses on the two endpoints of the MPTCP connection, i.e., the source and the destination. As a result, it cannot be applied directly to the user cooperation scenario, as a relay in the middle may have a great impact on the end-to-end performance.

This section first investigates the performance of the congestion control algorithm of MPTCP in a user cooperation scenario. In addition to providing the relaying service for the MPTCP connection between the source and the destination, the relays also have their local traffic based on TCP or more generic AIMD protocols [6]. Although TCP is the de facto standard for Internet applications, it is also meaningful to consider the more generic AIMD protocols, which can satisfy differentiated QoS requirements.

In AIMD-based congestion control, the source additively increases the congestion window by α units for each RTT and multiplicatively decreases the window size to a fraction β of its previous value whenever there is a congestion indication, e.g., triple duplicate acknowledgements. TCP is actually a special case of AIMD with $\alpha = 1$ and $\beta = 0.5$. On one hand, AIMD protocols are flexible to offer service differentiation by adapting the (α, β) pair. On the other hand, an AIMD flow can be ensured to be TCP-friendly if it satisfies

$$\alpha = \frac{3(1 - \beta)}{1 + \beta} \tag{7.2}$$

where $0 < \alpha < 3$ and $0 < \beta < 1$ [1].

Ideally, it is expected that, even when a relay is forwarding packets to the destination, the same throughput should be achieved for its local single-path

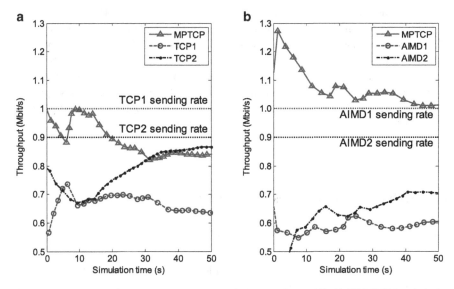

Fig. 7.1 Throughput of MPTCP flows and local single-path flows [11]. (**a**) With TCP local single-path flows. (**b**) With AIMD local single-path flows

traffic flow as that when there were no MPTCP subflows sharing the relay path. Any possible performance degradation can jeopardize the motivation for user cooperation. To demonstrate the effect of user cooperation on MPTCP performance, the user cooperation in Fig. 3.1 is considered for the LTE network. The source sends packets to the destination via two relays. Each relay runs a local single-path TCP or AIMD ($\alpha = 0.7$, $\beta = 0.625$) flow, while acting as a relay for a TCP subflow to the destination.

Figure 7.1 shows the throughput of all single-path TCP or AIMD flows and MPTCP flows with the original MPTCP congestion control algorithm [5]. The dotted straight lines represent the sending rates of the single-path flows, which are greater than half of the capacity of the LTE link between the eNB and each relay. As seen in Fig. 7.1a, the MPTCP aggregate throughput converges to the larger throughput of the two single-path TCP flows, which satisfies both objectives that the MPTCP congestion control algorithm claims [5]. Meanwhile, the throughput of single-path TCP flows becomes less than their corresponding sending rates, due to the bandwidth competition and sharing with the MPTCP subflows. Similar observations can be made in Fig. 7.1b, where the relays run single-path AIMD traffic flows locally. Moreover, the MPTCP aggregate throughput eventually converges to a level that significantly overwhelms the larger throughput of the two single-path AIMD flows, which violates the second objective of MPTCP. This is because the aggressiveness factor of MPTCP cannot guarantee no throughput degradation for single-path AIMD flows.

This problem is also harmful to the aggregate throughput of MPTCP with SSRS. In SSRS, the destination selects relays based on the available bandwidth of each relay. When the local traffic of the relays are controlled by TCP or AIMD protocols, the bandwidths of the MPTCP subflows observed at the destination can be quite different from the available bandwidths advertised by the relays, which may lead to unexpected outages for the achievable aggregate throughput. As a consequence, SSRS may not be able to guarantee a stable aggregate throughput with MPTCP for all the time, since the destination may update the active relay set too frequently due to the outage events.

7.2 Extensions to Congestion Control

7.2.1 Fairness for User Cooperation

As illustrated in Chap. 3, there can be multiple transmission paths from the source to the destination through different relays. The relays forward packets to the destination via Wi-Fi links, which have a higher transmission rate than the LTE links between the eNB and the relays. Hence, the LTE links can be assumed to be the bottleneck links of the paths. The *total bandwidth* (in Mbit/s), B_r, of an end-to-end path r via a relay n_r is then limited by the capacity of the bottleneck link between the eNB and n_r. The *available bandwidth* (in Mbit/s) of path r is defined as the unused bandwidth $B_r - C_r$, where C_r is the total traffic rate of flows over the bottleneck link. The available bandwidth may not be fully utilized due to factors such as a small receive buffer and flow control, which restrict the achieved end-to-end throughput [3].

Let K_r denote the number of local single-path flows over path r, and $\lambda_{r,l}$ denote the sending rate of a single-path flow f_l over path r. Assume $\lambda_{r,l} \leq \lambda_{r,j}$ if $l < j$ and $1 \leq l, j \leq K_r$. As such, the total traffic of local single-path flows over the path r becomes $\sum_{l=1}^{K_r} \lambda_{r,l}$. The available bandwidth of path r is then given by $B_r - \sum_{l=1}^{K_r} \lambda_{r,l}$, which is the maximum throughput that an additional multipath subflow aims to achieve without degrading the local flows.

Suppose that the sending rates of J_r ($J_r \leq K_r$) single-path local flows of relays are each less than $\xi_r = \frac{B_r}{K_r+1}$, which is the even share of the total bandwidth of path r, when the local flows further share the path with an additional multipath subflow. If the standard TCP and MPTCP congestion control were followed [5], the other $K_r - J_r$ TCP flows each having a sending rate greater than ξ_r could not approach their sending rates for the throughput due to the competition of the MPTCP subflow. As a result, the expected throughput of the MPTCP subflow becomes $(B_r - \sum_{l=1}^{J_r} \lambda_{r,l})/(K_r - J_r + 1)$. Hence, the ratio of the expected throughput of the MPTCP subflow based on the standard control to the maximum throughput without degrading local traffic is defined as

$$\rho_r = \frac{\frac{B_r - \sum_{l=1}^{J_r} \lambda_{r,l}}{K_r - J_r + 1}}{B_r - \sum_{l=1}^{K_r} \lambda_{r,l}} \tag{7.3}$$

For example, suppose that two local single-path TCP flows (A and B) run over a path r of a total bandwidth of 3 Mbit/s. The sending rates of A and B are 0.5 and 1.5 Mbit/s, respectively. If an MPTCP subflow of bulk data transfer further uses the path, fair sharing among all flows should lead to a throughput of 1 Mbit/s each. Since the sending rate of B (1.5 Mbit/s) is greater than the fair share (1 Mbit/s), the achieved throughput of B will be affected due to the bandwidth competition of the MPTCP subflow. In this example, $B_r = 3$ Mbit/s, $K_r = 2$, $J_r = 1$, $\sum_{l=1}^{K_r} \lambda_{r,l} = (0.5 + 1.5) = 2$ Mbit/s, $\sum_{l=1}^{J_r} \lambda_{r,l} = 0.5$ Mbit/s. Therefore, the expected throughput of the TCP flow B and MPTCP subflow will be $\frac{3 - 0.5}{2 - 1 + 1} = 1.25$ Mbit/s, which is the numerator of (7.3). The denominator of (7.3) gives the maximum throughput that the MPTCP subflow can achieve without degrading the throughput of A and B, which is $3 - (0.5 + 1.5) = 1$ Mbit/s. Hence, according to (7.3), $\rho_r = 1.25$. Here, $\rho_r > 1$ implies that the MPTCP subflow will take more than the unused bandwidth under the standard control and exceed the maximum throughput without degrading local traffic. Therefore, the increasing scale of the congestion window for the MPTCP subflow should be further constrained.

For generality, let (α_r, β_r) denote the increasing and decreasing parameters of the multipath subflow on path r and (α_r', β_r') for those of the local single-path flow. In the special case of MPTCP for the multipath subflow and TCP for the single-path flow, $\beta_r = 0.5$, $\alpha_r' = 1$, and $\beta_r' = 0.5$. It is known that the mean throughput of a flow in a steady state is proportional to its average congestion window size. Let $\overline{W}_{a,r}$ and $\overline{W}_{m,r}$ denote the average congestion window size of a single-path AIMD flow and that of the multipath subflow on path r, respectively. To satisfy the ratio in (7.3), it requires that [1]

$$\overline{W}_{a,r} = \rho_r \overline{W}_{m,r}. \tag{7.4}$$

Based on the analytical approach in [1], Appendix derives the following equations [8]:

$$\overline{W}_{a,r} = \frac{1 + \beta_r'}{2} \frac{\alpha_r'(1 - \beta_r)Y_r}{\tau}, \qquad \overline{W}_{m,r} = \frac{1 + \beta_r}{2} \frac{\alpha_r(1 - \beta_r')Y_r}{\tau} \tag{7.5}$$

where $\tau = \alpha_r' + \alpha_r - \alpha_r \beta_r' - \alpha_r' \beta_r$. When the total congestion window size of the single-path AIMD flow and the multipath subflow is no less than Y_r, it is referred to as an overload region. Then, (7.4) and (7.5) can be combined to obtain

$$\alpha_r = \frac{\alpha_r'(\beta_r' + 1)(\beta_r - 1)}{\rho_r(\beta_r' - 1)(\beta_r + 1)}. \tag{7.6}$$

To ensure TCP-friendliness, the (α'_r, β'_r) of the single-path flow should satisfy (7.2). That is, $\alpha'_r = \frac{3(1+\beta'_r)}{1-\beta'_r}$. Then, (7.6) is simplified to

$$\alpha_r = \frac{1}{\rho_r} \cdot \frac{3(1-\beta_r)}{1+\beta_r}. \tag{7.7}$$

When the multipath subflow is based on MPTCP, $\beta_r = 0.5$ and

$$\alpha_r = \frac{1}{\rho_r}. \tag{7.8}$$

For any generic multipath subflow with the increasing and decreasing parameters (α_r, β_r) on path r, given $\beta_r = \beta'_r$, it is further obtained that

$$\alpha_r = \frac{1}{\rho_r} \cdot \frac{3(1-\beta'_r)}{1+\beta'_r} = \frac{1}{\rho_r} \cdot \alpha'_r. \tag{7.9}$$

7.2.2 Extended Aggressiveness Factor

As observed in Fig. 7.1b, MPTCP cannot work well with local single-path AIMD flows because the MPTCP subflow is too aggressive and taking too much bandwidth of the LTE relay link. As a result, the throughput of local AIMD flows is significantly decreased. Therefore, when there is no bandwidth competition (i.e., $\rho \leq 1$) between the local single-path flow and the multipath subflow sharing the same path, the aggressive multipath subflow needs to be further regulated. The following shows how to extend the aggressiveness factor of MPTCP to regular the multipath subflow.

Referring to the objective (ii) of MPTCP, the multipath flow should take no more capacity than any single-path flow that shares the path when experiencing the same loss rate. The throughput of the multipath flow when subject to the same loss rate is upper bounded by the maximum throughput of the single-path flow, i.e.,

$$\sum_{r=1}^{K_s} \frac{w_{m,r}}{RTT_r} = \max_{1 \leq r \leq K_s} \frac{w_{a,r}}{RTT_r} \tag{7.10}$$

where $w_{m,r}$ and $w_{a,r}$ are the congestion window sizes of the multipath subflow and single-path AIMD flow on path r, respectively. The left-hand side of (7.10) gives the aggregate throughput of the multipath flow, while the right-hand side is the maximum throughput of the single-path AIMD flow.

For stability consideration, an increasing size of the congestion window for the multipath subflow on path r should be balanced with a decreasing amount in equilibrium. According to the congestion control algorithm of MPTCP, the source increases the congestion window of path r by $\min(a/w_{total}, 1/w_r)$ once an ACK is received from path r. This idea can be generalized to extend the aggressiveness factor a_r given a general increasing parameter α_r for path r. Hence,

$$(1 - p_r) \min \left(\frac{a_r}{w_{total}}, \frac{\alpha_r}{w_{m,r}} \right) = p_r w_{m,r} \beta_r \qquad (7.11)$$

where p_r is the packet loss rate (in percentage) of path r, w_{total} is the total congestion window size of the multipath subflows, (α_r, β_r) are the increasing and decreasing parameters of the multipath subflow on path r, and a_r is the aggressive factor for the multipath subflow on path r.

Likewise, the single-path AIMD flow is subject to the same path loss and satisfies

$$(1 - p_r) \frac{\alpha'_r}{w_{a,r}} = p_r w_{a,r} \beta'_r \qquad (7.12)$$

where (α'_r, β'_r) are the increasing and decreasing parameters of the local single-path AIMD flow. Then, (7.12) is rearranged to

$$\frac{p_r}{1 - p_r} = \frac{1}{(w_{a,r})^2} \frac{\alpha'_r}{\beta'_r}. \qquad (7.13)$$

To achieve the objective of regulating the aggressive multipath subflow, the aggressiveness factor a_r needs to guarantee $\frac{a_r}{w_{total}} \le \frac{\alpha_r}{w_{m,r}}$. Here, $\frac{a_r}{w_{total}}$ is the overall increasing size of the multipath flow, which also implies the achievable throughput of the multipath flow. In addition, $\frac{\alpha_r}{w_{m,r}}$ is the increasing size of the multipath subflow over path r. As the increasing parameter of the single-path flow over path r is unknown to the multipath subflow for congestion control, it is assumed that the increasing parameter of the single-path AIMD flow is the same as that of the multipath subflow sharing the same path. That is, $\alpha'_r = \alpha_r$. Hence, $\frac{\alpha_r}{w_{m,r}}$ can imply the achievable throughput of the single-path AIMD flow over path r. Therefore, $\frac{a_r}{w_{total}} \le \frac{\alpha_r}{w_{m,r}}$ ensures that the overall throughput of the multipath flow is no greater than the throughput of the single-path AIMD flow sharing the same path.

Thus, (7.13) is applied to (7.11) to have

$$a_r = w_{total} w_{m,r} \beta_r \frac{p_r}{1 - p_r}$$

$$= w_{total} w_{m,r} \beta_r \left[\frac{1}{(w_{a,r})^2} \frac{\alpha'_r}{\beta'_r} \right]$$

$$= \alpha'_r w_{total} \frac{w_{m,r}}{(w_{a,r})^2} \frac{\beta_r}{\beta'_r}.$$

Similar to Sect. 7.2.1, when $\beta_r = \beta'_r$, the aggressiveness factor a_r is rewritten as

$$a_r = \alpha'_r w_{total} \frac{w_{m,r}}{(w_{a,r})^2}. \qquad (7.14)$$

Considering the maximum throughput of the single-path AIMD flow in (7.10), a_r can be expressed in a form similar to (7.1) as follows:

$$a_r = \alpha'_r w_{total} \frac{\frac{w_{m,r}}{RTT_r^2}}{(\frac{w_{a,r}}{RTT_r})^2}$$

$$= \alpha'_r w_{total} \frac{\max_{1 \leq s \leq K_s} \frac{w_{m,s}}{RTT_s^2}}{\max_{1 \leq s \leq K_s} \left(\frac{w_{a,s}}{RTT_s}\right)^2}$$

$$= \alpha'_r w_{total} \frac{\max_{1 \leq s \leq K_s} \frac{w_{m,s}}{RTT_s^2}}{\sum_{s=1}^{K_s} \left(\frac{w_{m,s}}{RTT_s}\right)^2}. \tag{7.15}$$

Based on the new aggressiveness factor in (7.15), it is guaranteed that the multipath subflow does not take more bandwidth than a single-path flow when experiencing the same loss rate.

At last, it is worth mentioning that the packet loss rate p_r in (7.11) refers in particular to the loss due to congestion. In a wireless environment, packet loss can be induced by congestion as well as by transmission errors. On one hand, it is expected that the lower physical and link layers can address well packet loss due to transmission errors. On the other hand, the multipath congestion control algorithm can be properly integrated with a wireless TCP solution as a sublayer below. For example, in [4], a source of single-path TCP differentiates loss due to congestion or link errors based on explicit congestion notification (ECN). In the latter case, the source retransmits packets without decreasing the congestion window. Only if explicit congestion signal is received over a path should the source decrease its congestion window according to the multipath congestion control algorithm. As such, the packet loss occurring in a lower layer is hidden to the multipath transport layer.

7.2.3   Two Cases with Local TCP or AIMD Flows

Based on the analysis in Sects. 7.2.1 and 7.2.2, the MPTCP congestion control (MCC) algorithm is extended to ensure fair bandwidth sharing in the user cooperation scenario:

- Once the source receives an ACK message from path r, it increases its congestion window w_r for path r by $\min(a_r/w_{total}, \alpha'_r/w_r)$ if $\rho_r \leq 1$, or by $\min(\alpha'_r/w_r, \alpha'_r/(\rho_r w_r))$ if $\rho_r > 1$; and
- Once the source receives a congestion signal from path r, it decreases its congestion window for path r to $\beta_r w_r$.

Since $\rho_r > 1$ implies that the multipath subflow is taking more than the unused bandwidth and degrading local traffic, the increasing scale of the congestion window for the multipath subflow is limited by $\min(\alpha'_r/w_r, \alpha'_r/(\rho_r w_r))$. The first parameter α'_r/w_r restricts the multipath subflow so that it does not increase its congestion window faster than the single-path AIMD flow. Here, α'_r is the increasing parameter of the single-path AIMD flow sharing path r. The second parameter $\alpha'_r/(\rho_r w_r)$ is based on the condition in (7.9).

On the other hand, when $\rho_r \leq 1$, there is no bandwidth competition between the multipath subflow and local single-path flow. In this case, the congestion window increasing parameter for the multipath subflow is defined as $\min(a_r/w_{total}, \alpha'_r/w_r)$. The first parameter a_r/w_{total} regulates the aggressive behavior of the multipath subflow, so that the multipath flow achieves an aggregate throughput no greater than that of the single-path flow. Here, a_r is the extended aggressiveness factor given in (7.14), which is also expressed in another form in (7.15). Similar to the case with $\rho_r > 1$, the second parameter α'_r/w_r ensures fair sharing between the multipath subflow and the single-path AIMD flow.

It is worth noting that the extended congestion control algorithm requires that the multipath subflow obtain α'_r through additional signalling. In practice, a well-known fixed (α, β) pair is often defined for a specific scenario in advance [6]. Also, ρ_r can be measured by the destination and signalled back to the source in a certain frequency, e.g., for every 2 s in the experiments in Sect. 7.3.

Based on the above algorithm, there are two extensions to MPTCP in a user cooperation scenario when the local single-path flow are based on TCP or generic AIMD, referred to as *MCC-Coop* and *GMCC-Coop*, respectively. The following specifies the first extension *MCC-Coop*:

- Once the source receives an ACK from path r, it increases the congestion window of path r by $\min(a/w_{total}, 1/w_r)$ if $\rho_r \leq 1$, or by $\min(1/w_r, 1/(\rho_r w_r))$ if $\rho_r > 1$; and
- Once the source receives a congestion signal from path r, it decreases the congestion window w_r of path r to $w_r/2$.

Note that $\alpha'_r = 1$ and $\beta'_r = 0.5$ for the single-path TCP flow, while $\beta_r = 0.5$ for MPTCP. Hence, the aggressiveness factor defined in (7.15) reverts to the original aggressiveness factor of MPTCP given in (7.1).

The more generic extension *GMCC-Coop* for MPTCP with the local single-path AIMD flow is described as follows:

- Once the source receives an ACK message from path r, it increases its congestion window w_r for path r by $\min(a_r/w_{total}, \alpha'_r/w_r)$ if $\rho_r \leq 1$, or by $\min(\alpha'_r/w_r, \alpha'_r/(\rho_r w_r))$ if $\rho_r > 1$; and
- Once the source receives a congestion signal from path r, it decreases its congestion window for path r to $w_r/2$.

7.3   Experimental Results of Bandwidth Sharing Algorithms[1]

This section evaluates the aggregate throughput of MPTCP with MCC-Coop or GMCC-Coop in both the static and dynamic scenarios. The local traffic at the relays is controlled by TCP or AIMD protocols. In the experiments, the ratio ρ_r defined in (7.3) is measured in every 2 s at the destination. Also, the same system topology with two fixed relays as defined in Chap. 3 is considered. The other simulation parameters are the same as the default parameters given in Table 3.1.

7.3.1   Static Scenario

The performance of MCC-Coop and GMCC-Coop is first evaluated in a static scenario, where the destination is connected to two fixed relays. Each relay forwards packets for one MPTCP subflow between the source and the destination.

Figure 7.2 shows the aggregate throughput of the MPTCP flow for the destination, and the individual throughput of local single-path flows of the relays. In Fig. 7.2a, the sending rates of the two single-path TCP flows over the two relays are 1.0 and 0.9 Mbit/s, respectively. Because the total bandwidth of each relay is 1.2 Mbit/s, $\rho_1 = 5$ and $\rho_2 = 3$, both greater than 1. Hence, MCC-Coop is activated for both paths to protect the local traffic of the relays. As seen in Fig. 7.2a, the throughput of the local TCP flows converges to their corresponding sending rates. This exactly achieves the design objective that the throughput of local TCP flows is not degraded by the MPTCP subflow sharing the same path. A similar observation for MPTCP with GMCC-Coop can be made from Fig. 7.2b, where the local traffic consists of two single-path AIMD flows with $\alpha'_r = 0.7$ and $\beta'_r = 0.625$.

7.3.2   Dynamic Scenario

This section evaluates MCC-Coop and GMCC-Coop in a more dynamic scenario. The SSRS algorithm in Chap. 4 is used to guarantee a stable aggregate throughput of MPTCP to meet the target bit rate (TBR) requirement at the destination. In SSRS, the destination dynamically maintains an active relay set, whose total available bandwidth falls into a range $[(1-\theta)\text{TBR}, (1+\theta)\text{TBR}]$, $0 < \theta < 1$. In the following experiments, the TBR and θ are set to 3 Mbit/s and 10 %, respectively.

In the dynamic scenario, the sending rates of the local traffic of relays are time-varying during the simulation. Moreover, in order to examine how MCC-Coop and

---

[1]Reprinted with permission from IEEE Network (2014), "Multipath TCP for user cooperation in LTE networks," by D. Zhou, W. Song, P. Wang, and W. Zhuang [11].

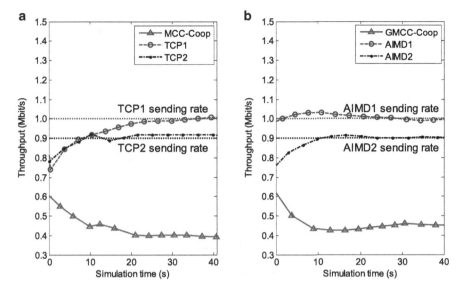

Fig. 7.2 Throughput of MPTCP flows with MCC-Coop and GMCC-Coop and local single-path flows in the static scenario. (**a**) With TCP local single-path flows. (**b**) With AIMD local single-path flows

GMCC-Coop respect the local traffic of the relays and help SSRS to achieve a stable aggregate throughput, the dynamic scenario in the simulation is divided into three special stages:

- S_1 for the period from the 10th s to the 20th s: the sending rate of *each relay* in the active set is *less* than half of the total bandwidth on the corresponding path via that relay;
- S_2 for the period from the 20th s to the 40th s: the sending rate of one relay in the active set is *less* than half of the total bandwidth of the corresponding path via that relay, while the sending rate of the other relay in the active set is *greater* than half of the total bandwidth of the corresponding path via that relay; and
- S_3 for the period from the 40th s to the 60th s: the sending rate of *each relay* in the active set is *greater* than half of the total bandwidth on the corresponding path via that relay.

Figure 7.3 compares the throughput of MPTCP and the extension MCC-Coop. In order to demonstrate performance variations with network dynamics, average throughput for every 2 s is considered. As seen in Fig. 7.3, because the sending rates of all active relays in S_1 are very low (e.g., < 0.2 Mbit/s), only the standard congestion control algorithm of MPTCP takes effect. Therefore, the aggregate throughput of MPTCP with MCC-Coop is very close to that of the standard MPTCP congestion control. As the local traffic in some relays is increased in S_2, a new active set with three relays is selected by SSRS to satisfy the TBR requirement. The sending rate of one particular relay is larger than half of the total bandwidth on the

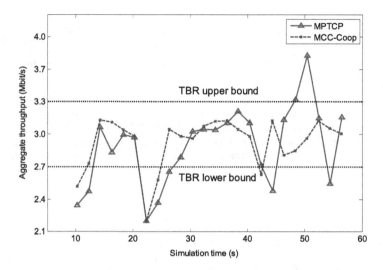

Fig. 7.3 Aggregate throughput of MPTCP and MCC-Coop with SSRS in the dynamic scenario

path via this relay. As already shown in Fig. 7.1a, the standard congestion control of MPTCP tends to take more bandwidth aggressively and thus harms the local traffic in that relay. At the end of S_2, the aggregate throughput of MPTCP is greater than that of MCC-Coop.

The aggressive behavior of MPTCP becomes more evident in S_3 when all active relays are sending their local TCP traffic at rates greater than half of the path bandwidths. As a result, the aggregate throughput of MPTCP exceeds the TBR upper bound, which triggers SSRS to update the active relay set. Consequently, the aggregate throughput of MPTCP fluctuates at a much larger scale, while the aggregate throughput of MCC-Coop is more stable and stays within the TBR range.

The GMCC-Coop extension in Sect. 7.2.2 addresses particularly the bandwidth sharing between MPTCP subflows and local AIMD flows at relays. Depending on the projected throughput ratio ρ_r, the congestion window of the MPTCP subflow is increased in different manners. To illustrate the effect of this differentiation, a reference algorithm, referred to as *MPAIMD* [8], is considered. Following an ACK reception, MPAIMD always increases the congestion window w_r for path r by $\min(a_r/w_{total}, \alpha'_r/w_r)$. That is, MPAIMD does not distinguish different cases of ρ_r by always assuming $\rho_r \leq 1$ and thus neglects the bandwidth competition and potential harm to local single-path traffic.

Figure 7.4 compares the aggregate throughput of GMCC-Coop and MPAIMD with local AIMD flows. Similar to Fig. 7.3, it is observed that GMCC-Coop is not only less aggressive in S_2 and S_3 when there are higher local traffic demands, but also provides a more stable aggregate throughput than MPAIMD.

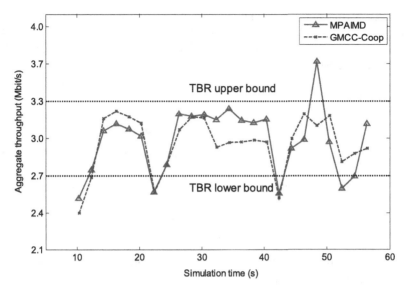

Fig. 7.4 Aggregate throughput of MPAIMD and GMCC-Coop with SSRS in a dynamic scenario

7.4 Summary

This chapter focuses on the bandwidth sharing between the MPTCP subflows and the local single-path flows of relays for a user cooperation scenario in the LTE network. Two types of local traffic at the relays are considered, including TCP and AIMD single-path flows. The extensions, MCC-Coop and GMCC-Coop, aim to restrict the aggressive behavior of MPTCP and protect the local TCP and AIMD traffic at the relays. Specifically, the MPTCP subflow on a path cannot exceed a maximum throughput, which is the available bandwidth of the relay on the path so as not to degrade the throughput of local traffic at the relays. Depending on the ratio of the expected throughput under the standard MPTCP congestion control to the maximum throughput allowed for the MPTCP subflow, the congestion window of the MPTCP subflow, especially the increasing scale, is adapted in different manners. Simulation results demonstrate that the solutions are effective in both the static and dynamic scenarios. It is shown that MCC-Coop and GMCC-Coop not only guarantee the throughput of local traffic at the relays, but also ensure that SSRS achieves a stable aggregate throughput to satisfy the TBR requirement.

Appendix: Proof of Equation (7.5)

This appendix presents the derivation of (7.5) using the analytical method in [1]. Consider a scenario where a single-path AIMD flow and a multipath subflow share a common path r. Let (α_r, β_r) denote the general increasing and decreasing parameters of the multipath subflow on path r and (α'_r, β'_r) represent those of the local single-path AIMD flow. Let $W_{a,r}(t)$ and $W_{m,r}(t)$ denote the congestion window size at time t of a single-path AIMD flow and that of the multipath subflow on path r, respectively.

According to Eqs. (2) and (3) in [1], the following equations are obtained for the above scenario:

$$W_{a,r}(t + \Delta t) = W_{a,r}(t) + \alpha'_r \Delta t \tag{7.16}$$

$$W_{m,r}(t + \Delta t) = W_{m,r}(t) + \alpha_r \Delta t \tag{7.17}$$

where Δt is a short time period. Then, (7.16) and (7.17) can be combined to have

$$\frac{W_{m,r}(t + \Delta t) - W_{m,r}(t)}{W_{a,r}(t + \Delta t) - W_{a,r}(t)} = \frac{\alpha_r}{\alpha'_r}. \tag{7.18}$$

This implies that the slope of the congestion window size of the single-path AIMD flow versus that of the multipath subflow on path r is α_r / α'_r (see Fig. 1 of [1]).

Consider an overload region for path r when the total congestion window size of the single-path AIMD flow and the multipath subflow is no less than Y_r. Then, Eqs. (5)–(7) in [1] can be rewritten as

$$W_{a,r}(t_l) + W_{m,r}(t_l) = Y_r \tag{7.19}$$

$$W_{a,r}(t_l^+) = \beta'_r W_{a,r}(t_l) \tag{7.20}$$

$$W_{m,r}(t_l^+) = \beta_r W_{m,r}(t_l) \tag{7.21}$$

where t_l is the time that two flows enter into the overload region for the lth time, where $l \geq 1$. Assume that both flows receive the congestion signal once their total sending rates exceed the link capacity, and decrease their congestion window simultaneously. Then, $W_{a,r}(t_l^+)$ and $W_{m,r}(t_l^+)$ denote the immediate decreased congestion window size of the single-path AIMD flow and that of the MPTCP subflow, respectively.

Based on (7.18)–(7.21), it is further proved that

$$
\begin{aligned}
&\frac{W_{m,r}(t_{l+1}) - W_{m,r}(t_l^+)}{W_{a,r}(t_{l+1}) - W_{a,r}(t_l^+)} \\
&= \frac{W_{m,r}(t_{l+1}) - \beta_r W_{m,r}(t_l)}{W_{a,r}(t_{l+1}) - \beta'_r W_{a,r}(t_l)} \\
&= \frac{W_{m,r}(t_{l+1}) - \beta_r W_{m,r}(t_l)}{(Y_r - W_{m,r}(t_{l+1})) - \beta'_r (Y_r - W_{m,r}(t_l))} \\
&= \frac{\alpha_r}{\alpha'_r}.
\end{aligned}
\tag{7.22}
$$

Then, $W_{m,r}(t_{l+1})$ can be expressed as

$$W_{m,r}(t_{l+1}) = \frac{\alpha_r \beta_r' + \alpha_r' \beta_r}{\alpha_r + \alpha_r'} W_{m,r}(t_l) + \ + \frac{\alpha_r(1 - \beta_r')}{\alpha_r + \alpha_r'} Y_r. \tag{7.23}$$

Here, (7.23) can be rearranged to

$$W_{m,r}(t_{l+1}) - W_{m,r}(t_l) = -\frac{\alpha_r(1 - \beta_r') + \alpha_r'(1 - \beta_r)}{\alpha_r + \alpha_r'} W_{m,r}(t_l) + \frac{\alpha_r(1 - \beta_r')}{\alpha_r + \alpha_r'} Y_r. \tag{7.24}$$

Following the analysis in [1], when $W_{m,r}(t_{l+1}) - W_{m,r}(t_l) \to 0$ to reach the steady state, the congestion window $W_{m,r}(t_l)$ converges to $\alpha_r(1 - \beta_r')Y_r/\tau$, where $\tau = \alpha_r' + \alpha_r - \alpha_r \beta_r' - \alpha_r' \beta_r$. Similarly, $W_{a,r}(t_l)$ converges to $\alpha_r'(1 - \beta_r)Y_r/\tau$. Then, based on Eqs. (11) and (12) in [1], the average congestion window size can be calculated by

$$\overline{W}_{a,r} = \frac{1 + \beta_r'}{2} \frac{\alpha_r'(1 - \beta_r)Y_r}{\tau}, \qquad \overline{W}_{m,r} = \frac{1 + \beta_r}{2} \frac{\alpha_r(1 - \beta_r')Y_r}{\tau}. \tag{7.25}$$

References

1. Cai, L., Shen, X., Pan, J., Mark, J.W.: Performance analysis ot TCP-friendly AIMD algorithms for multimedia applications. IEEE Trans. Multimedia 7(2), 339–355 (2005)
2. Ford, A., Raiciu, C., Handley, M., Barre, S., Iyengar, J.: Architectural guidelines for multipath TCP development. IETF RFC 6182 (2011)
3. Hu, N., Steenkiste, P.: Evaluation and characterization of available bandwidth probing techniques. IEEE J. Select. Areas Commun. 21(6), 879–894 (2003)
4. Liu, J., Singh, S.: ATCP: TCP for mobile ad hoc networks. IEEE J. Select. Areas Commun. 19(7), 1300–1315 (2001)
5. Raiciu, C., Handley, M., Wischik, D.: Coupled congestion control for multipath transport protocols. IETF RFC 6356 (2011)
6. Yang, Y., Lam, S.S.: General AIMD congestion control. In: Proc. IEEE ICNP (2000)
7. Zhou, D., Ju, P., Song, W.: Performance enhancement of multipath TCP with cooperative relays in a collaborative community. In: Proc. IEEE PIMRC (2012)
8. Zhou, D., Song, W., Cheng, Y.: A study of fair bandwidth sharing with AIMD-based multipath congestion control. IEEE Wireless Communications Letters 2(3), 299–302 (2013)
9. Zhou, D., Song, W., Ju, P.: Subset-sum based relay selection for multipath TCP in cooperative LTE networks. In: Proc. IEEE GLOBECOM (2013)
10. Zhou, D., Song, W., Shi, M.: Goodput improvement for multipath TCP by congestion window adaptation in multi-radio devices. In: Proc. IEEE CCNC (2013)
11. Zhou, D., Song, W., Wang, P., Zhuang, W.: Multipath TCP for user cooperation in LTE networks. IEEE Network. http://cs.unb.ca/~wsong/publications/journals/NET_MPTCP_Dizhi.pdf (2014)

Chapter 8
Conclusions and Future Directions

8.1 Conclusions

This brief focuses on efficient multipath transmission based on MPTCP [2] with user cooperation in the LTE network. In a user cooperation scenario, nearby relays can receive packets on behalf of the destination via their own LTE links and then forward the packets toward the destination via Wi-Fi links. By this means, the destination can aggregate the available bandwidth of the relays by enabling an MPTCP connection, which includes multiple subflows from the source to the destination through different relays.

Nevertheless, in order to provide a stable QoS to the upper layers, MPTCP needs to address several critical issues: (1) how does MPTCP offer a stable aggregate throughput to the application at the destination in a highly dynamic user cooperation scenario? (2) how does MPTCP guarantee a stable goodput, which reflects the real application-level throughput, based on the steady aggregate throughput? and (3) how does MPTCP respect the local traffic of the relays? This means that the relay should be guaranteed the same throughput for the local traffic as that when it does not forward the traffic for the destination, which can be seen as a key motivation for mobile users to provide any relaying service.

In order to address the above challenges, this brief presents several state-of-the-art solutions, which are reviewed in the following.

- Subset-sum based relay selection (SSRS) [4], which is an application layer enhancement module for MPTCP and runs at the relay and the destination. SSRS aims to guarantee a stable aggregate throughput to satisfy the TBR requirement of the specific application at the destination. Based on a fully polynomial-time relay selection algorithm, SSRS efficiently selects multiple relay sets whose total available bandwidths are within an acceptable TBR range, e.g., between 90 and 110 % of TBR. Then, SSRS chooses the active and backup relay sets based on several criteria, e.g., the available bandwidths and the number of relays.

© The Author(s) 2014
D. Zhou, W. Song, *Multipath TCP for User Cooperation in Wireless Networks*,
SpringerBriefs in Computer Science, DOI 10.1007/978-3-319-11701-0_8

Once the aggregate throughput is observed out of the TBR range at the destination, the active set is smoothly migrated to the backup set. The backup set is updated periodically at the destination by monitoring the available bandwidth variations at the relays. The simulation results demonstrate that SSRS achieves a stable aggregate throughput in both the static and dynamic available bandwidth patterns at the relays.

- Adaptive congestion control (ACC) [5], which is a module based on the standard MPTCP congestion control at the source and aims to enhance the goodput performance of MPTCP. Although SSRS can guarantee a stable aggregate throughput with MPTCP, the goodput observed at the application layer of the destination is still far lower than the aggregate throughput, since the end-to-end delay differences of the relay paths can cause out-of-order packets. As the goodput is inversely proportional to the end-to-end path delay difference, ACC improves the goodput by dynamical adapting the congestion window of MPTCP subflows so as to minimize the path delay differences. Experimental results show that the goodput of MPTCP is significantly improved by using ACC together with SSRS. On average, the goodput of SSRS + ACC is 1.5 higher than that of SSRS alone in the static pattern, while the goodput of SSRS + ACC is 1.33 times higher than that of SSRS alone in the dynamic pattern.

- Differentiated packet forwarding (DPF) [3], which is another module that complements with ACC so as to further improve the goodput of MPTCP at the destination. In a user cooperation scenario, ACC cannot eliminate the delay differences of the paths, since the end-to-end path delay is affected by many other factors, which are not controlled by the endpoints, e.g., the queue length at the routers. DPF works in a distributed manner at the relays and the destination. The destination periodically sends an expected DSN range to the relays. The relays buffer the out-of-order packets outside this range and only forward the packets within this range to the destination. Further, a fast ACK scheme can be implemented at the relays to enable the relays to return ACK messages to the source for the buffered packets on behalf of the destination. This fast ACK scheme can reduce the delayed ACKs that cause unnecessary decrease of congestion window at the source. It is seen in the experiments results that the goodput of DPF is 1.32 times higher and 1.44 times higher than that of SSRS alone in static pattern and dynamic pattern, respectively. And the goodput achieved by DPF is more stable than that of ACC. It is also observed that the three modules (SSRS, ACC and DPF) can work well together and provide the highest and most stable goodput in both the static and dynamic available bandwidth patterns.

- The bandwidth sharing module [6] aims to guarantee fair bandwidth sharing between the MPTCP subflows and the local single-path flows at the relays. Two solutions, MCC-Coop and GMCC-Coop, which extend the standard MPTCP congestion control, are introduced. They restrict the aggressive behavior of MPTCP and protect the local TCP and AIMD traffic at the relays by regulating the increase of congestion window of each path based on a throughput ratio. As such, the throughput of the MPTCP subflow on a path is bound by a maximum throughput, which is the available bandwidth of the relay path. Also, a new aggressiveness factor is derived in GMCC-Coop so as to protect the local AIMD single-path traffic at the relays. Experimental results demonstrate that MCC-Coop and GMCC-Coop can guarantee that the throughput of local TCP and AIMD flows at the relays is not degraded in various scenarios.

8.2 Future Extensions

As discussed in Sect. 8.1, these extension modules can provide a stable aggregate throughput, greatly improve the goodput of MPTCP and well respect the local traffic at relays. In order to optimize the performance of MPTCP in a user cooperation scenario of the LTE network, the research can be further extended in the following aspects.

8.2.1 Advanced Traffic Models

This brief uses various traffic models to evaluate the performance of the extension modules. The bulk data model is used for MPTCP flow between the source and the destination, while the on-off model is used for single-path flows of relays to generate the background traffic in different patterns, e.g., the static pattern and the dynamic patterns.

In the future, traffic models for specific applications can be used so as to investigate the performance of the extension modules in more complex environments. On one hand, a video traffic model can be used to generate the local traffic at the relays so as to simulate a highly dynamic available bandwidth pattern [1]. On the other hand, the video traffic model can be considered for the MPTCP connection as well. Both cases may force the destination to re-select the relay set based on the available bandwidths of the relays or the TBR requirement of the destination. Although SSRS can efficiently update the relay set in these two scenarios, it still needs to further study how seriously these complex traffic models can affect the stability of the aggregate throughput of MPTCP.

8.2.2 Extensions to Existing Modules

Several potential extensions can be considered for each reviewed module to further improve their performance in more dynamic scenarios.

First, the fixed TBR variation range of aggregate throughput in the SSRS module can be extended to a dynamic range, which may vary with several criteria, such as the number of buffered packets at the destination and the network conditions.

In the case of a low aggregate throughput, the TBR variation range can be expanded to avoid handing over the MPTCP subflows to the backup relay set if there are enough packets buffered at the destination. As such, the application at the destination can retrieve the packets from the local buffer first so that the low aggregate throughput will not be perceived by the user immediately. Also, the relay selection algorithm of SSRS can be extended by taking into account more factors to achieve various additional objectives, e.g., to minimize the energy consumption by considering the energy cost of the relays.

Second, ACC can use a different maximum adaptation limit for each MPTCP subflow to control the congestion on each path more precisely. In ACC, the source uses the same maximum adaptation limit to avoid severe throughput degradation on each path. In fact, the MPTCP subflow with a large congestion window ($cwnd$) can stand more adaptations than the subflow with a smaller $cwnd$. Hence, it is worth further studying how to relate the maximum adaptation limit of a path with its congestion window size.

Third, in DPF, the destination can assign a different expected DSN range to each relay by considering additional factors, such as the buffer size of the relays. If a relay has a limited size of buffer for forwarding packets to the destination, the DSN range for the relay should be updated adaptively to avoid overflowing the buffer. The DSN range of the relays can also be determined by jointly considering the throughput and the delay difference of the MPTCP subflows. A relay path with a high throughput and a significant delay difference with other paths requires a larger DSN range to buffer the out-of-order packets. The destination can combine these factors together to decide the DSN range for each relay.

References

1. Dai, M., Zhang, Y., Loguinov, D.: A unified traffic model for MPEG-4 and H.264 video traces. IEEE Trans. Multi. **11**(5), 1010–1023 (2009)
2. Ford, A., Raiciu, C., Handley, M., Barre, S., Iyengar, J.: Architectural guidelines for multipath TCP development. IETF RFC 6182 (2011)
3. Zhou, D., Ju, P., Song, W.: Performance enhancement of multipath TCP with cooperative relays in a collaborative community. In: Proc. IEEE PIMRC (2012)
4. Zhou, D., Song, W., Ju, P.: Subset-sum based relay selection for multipath TCP in cooperative LTE networks. In: Proc. IEEE GLOBECOM (2013)
5. Zhou, D., Song, W., Shi, M.: Goodput improvement for multipath TCP by congestion window adaptation in multi-radio devices. In: Proc. IEEE CCNC (2013)
6. Zhou, D., Song, W., Wang, P., Zhuang, W.: Multipath TCP for user cooperation in LTE networks. IEEE Network. http://cs.unb.ca/~wsong/publications/journals/NET_MPTCP_Dizhi.pdf (2014)